THE UFO EXPERIENCE

THE UFO EXPERIENCE

Evidence Behind Close Encounters, Project Blue Book, and the Search for Answers

J. ALLEN HYNEK

This edition first published in 2023 by MUFON, an imprint of
Red Wheel/Weiser, LLC
With offices at:
65 Parker Street, Suite 7
Newburyport, MA 01950
www.redwheelweiser.com

 Originally published in Great Britain as *The UFO Experience* in 1972 by Abelard-Schuman.

ISBN: 978-1-59003-308-1

Library of Congress Cataloging-in-Publication Data

Names: Hynek, J. Allen (Joseph Allen), 1910–1986, author.
Title: The UFO experience : evidence behind close encounters, project blue book, and the search for answers / J. Allen Hynek.
Description: Newburyport, MA : MUFON, an imprint of Red Wheel/Weiser, LLC, 2023. | "Copyright 1972, 1975, 2023 by the Estate of J. Allen Hynek." | Includes bibliographical references and index. | Summary: "Cited by The New York Review of Books as "the best brief for visitation," this classic study presents an analysis of UFO reports and concludes that many sightings cannot be easily dismissed. The case against UFOs and UAPs has not been put to rest. Although UFOs "officially" did not exist for decades according to the government, reports of sightings continue to be made, and the latest releases from the government and related hearings have surprised the world. In The UFO Experience, a scientist of impeccable qualifications
takes on his colleagues"— Provided by publisher.
Identifiers: LCCN 2022057706 | ISBN 9781590033081 (trade paperback) | ISBN 9781633412903 (kindle edition)
Subjects: LCSH: Unidentified flying objects. | BISAC: BODY, MIND & SPIRIT / UFOs & Extraterrestrials | SCIENCE / Space Science / Astronomy
Classification: LCC TL789 .H9 2023 | DDC 001.942—dc23/eng/20221223
LC record available at https://lccn.loc.gov/2022057706

Cover design by Sky Peck Design
Interior photos courtesy of the Hynek Estate
Interior by Happenstance Type-O-Rama
Typeset in Adobe Caslon Pro, Almonte, and ITC Franklin Gothic Std

Printed in the United States of America
IBI
10 9 8 7 6 5 4 3 2 1

Contents

Foreword

I'm quite sure that when my father was asked by the Air Force's Project Blue Book, not long after World War II, to help them investigate UFOs, he thought it would be all of a few weekends' worth of work.

When he jumped into the unidentified deep end, hardly any other serious scientists would give flying saucers the time of day. This was the era of little green men with ray guns. It was not exactly what a tenure track professor would think of as a solid move. But jump he did.

By the time they were called UFOs, more and more scientists and influential figures started taking the phenomenon seriously and admitted so in public. Others, such as Neil Armstrong and Arthur C. Clarke, confided their alignment to my father's views in private. There was a recent article in the *Sun* talking about Neil Armstrong and my Dad.*

Nowadays, the government calls them UAPs—Unidentified Aerial Phenomena. They admit openly that they see them frequently in the air and in the water, and that they are concerned for both the safety of pilots and national security. A recent Pew poll showed that a majority of Americans believe aliens have visited Earth.

My father always wanted to push the jagged edges of mainstream science a bit further out. As is often the case, the way he did it was not what he originally had in mind. But expanding the limits of serious scientific discourse to include visitors from other planets or dimensions may well be the biggest gain of acceptable intellectual ground of all time.

---

* "Close Encounter: Astronaut Neil Armstrong Was 'Fascinated by UFOs, Secretly Met with Expert—and Even Tried to Protect Government Whistleblowers'," *The Sun*, January 25, 2019, *thesun.co.uk*.

Dr. Hynek and Neil Armstrong from a cruise in 1973

My father was uniquely well-positioned to push this cosmic envelope. He was a PhD astrophysicist from the University of Chicago who had worked on the proximity fuse during the war, and who would soon go on to install a dozen observatories around the world and organize the world's first international citizen science network to monitor the new dawn of artificial satellites. Many people identified with how he started off as a confirmed skeptic regarding UFOs, and went on to become not a believer, as that's not a word that scientists particularly fancy, but rather an accepter of the accumulated weight of the data.

And for experiencers, when this avuncular goateed professor with impeccable credentials said that not only was there a bona fide phenomenon that merited serious study, but that he also personally thought some experiencers sincerely felt that they had had some kind of genuine encounter, this helped many of them feel legitimized in thinking that maybe something extraordinary had actually happened, and also in disclosing their account to loved ones and strangers alike.

Dr. Hynek on the set of Steven Spielberg's *Close Encounters of the Third Kind* where he was an advisor on the film and made a cameo

UFOs have shown themselves to be a timeless phenomenon, and this book is no less relevant now than the day it was published. By reading it, you'll have the same insight that my father had—we aren't studying UFOs, we're studying UFO reports. That's why we needed the Close Encounters classification system—frameworks such as these are the building materials that science uses to create what are initially remote settlements beyond established beliefs, and which, in turn, become base camps for even further exploration.

So now, armed with your copy of *The UFO Experience*, buckle yourself in for a fascinating evidence-fueled journey to the new fringes of mainstream science.

—PAUL HYNEK

Preface

During my many years as scientific consultant to the United States Air Force on the matter of Unidentified Flying Objects I was often asked (and frequently still am) to recommend "a good book about UFOs." Very often, too, the request was accompanied by remarks along the line of "Is there really anything to this business at all?" "Just what's it all about anyway—is there any reliable evidence about UFOs?" or "Where can I read something about the subject that wasn't written by a nut?" With a few notable exceptions I have been hard-pressed to give a good answer to such questions. There are, of course, many books dealing with the subject. They regale the reader with one UFO story after another, each more spectacular than the other, but little space is devoted to documentation and to evaluation. What were the full circumstances surrounding the reported event? How reliable and how consistent were the reporters (all too often it is the lone reporter) of the event? And how were the UFO accounts selected? Most often one finds random accounts, disjointed and told in journalese.

I hope that this is a book to answer the questions of the person who is curious about the UFO phenomenon as a whole, who would like to have it appraised and to appraise it themselves.

I have often asked myself what "a good book on UFOs" would be like. Who would be qualified to write it, what should it contain, and what questions should it attempt to answer? I decided to try to write such a book, basing it on my twenty years of close association with the subject, during which time I had interrogated many hundreds of persons and personally investigated nearly as many cases. I decided to describe, primarily for the benefit of those who have been honestly puzzled by the UFO question, what UFO reports are like firsthand, what kind of people make them, what sorts of things the reports have in common, and how the subject has been presented and treated (I cannot honestly say "studied") in the past.

I cannot presume to describe, however, what UFOs *are* because I don't know; but I can establish beyond reasonable doubt that they are not *all* misperceptions or hoaxes. Indeed, those reports that do stem from *identifiable* sources do not, obviously, satisfy the definition of an *Unidentified* Flying Object. Misperceptions of aircraft, high-altitude balloons, meteors, and twinkling stars do account for many initial reports, but these do not qualify as UFO reports and need be treated only briefly in a book about UFOs. "A good book on UFOs" should, I think, be honest, without prejudgment; it should be factual and as well documented as possible. It should not be, however, a book that retails—or retells—UFO stories for the sake of their story value; rather it should attempt to portray the kinds of things that people—real everyday human beings with jobs and families—say they have actually experienced. These people are not merely names in a telephone book; they are flesh and blood persons who, as far as they are concerned, have had experiences as real to them as seeing a car coming down the street is to others.

I hope this book is one that will be recommended to you as "a good book on UFOs."

J. ALLEN HYNEK
NORTHWESTERN UNIVERSITY
EVANSTON, ILLINOIS
JANUARY 1, 1972

Acknowledgments

I am beholden to many persons for helping me in the preparation of this book. In particular, I wish to thank Professor Thomas Goudge, Department of Philosophy, University of Toronto, for his many helpful discussions of the subject matter in the early stages of preparation and Dr. Paul Davies, Institute of Theoretical Astronomy, University of Cambridge, England, for productive discussions in the latter stages. Dr. Harry Wood, Arizona State University, Ms. Jennie Zeidman, and Ms. Necia Apfel critically read the manuscript, and to their valuable suggestions were added those of Dr. Jacques Vallee, Stanford University, and Mr. Fred Beckman, of the University of Chicago.

I am indebted to Mr. William Weitzel, Ms. Josephine Clark, Mr. Ted Phillips, Mr. Warren Smith, Mr. Raymond Fowler, Mr. Bud Ledwith, and Ms. Isabel Davis for the use of material relating to UFO cases in this country and to Mr. W. K. Allan and Mr. Brian Cannon, Canadian UFO investigators, for making available Canadian UFO cases.

I am especially grateful to Ms. Mary Lou Armstrong for her permission to publish her letter of resignation as administrative assistant to Dr. Condon and to Mr. William Powers for permission to use portions of his critique of the Condon Report, which was refused publication in *Science.*

My grateful thanks go to all of the above and certainly to my secretary, Ms. Ann Larson, for her efficient and repeated typings of the manuscript.

Prologue

There is a sense in which each age is ripe for breakthroughs, for changes that were not only impossible but even frightening when imagined in an earlier age. Yet despite our potential for discovery, there is inherent in each epoch of our history a certain smugness that seems not to be apparent to most participants in that age. It is a complacent unawareness of the scope of things not yet known that later epochs look back upon with a sympathetic smile of condescension, if not with polite laughter.

By the same token, the breakthroughs and world concepts of the *future* probably would be unthinkable and certainly bewildering if we could now glimpse them. Yet changes in their proper time do occur, and it therefore behooves us to study seriously, not dismiss with scathing ridicule, the puzzling phenomena of today in the hope of coming upon satisfactory explanations. We may thus venture into the future, so to speak.

The UFO phenomenon may well be one such challenging area of interest even though it is seemingly out of place in our present world picture—as incredible to us as television would have been to Plato. The study of this frequently reported phenomenon may offer us an enticing glimpse of and point a beckoning finger to the future.

Occasionally scientists sense the presence of the intangible, awesome domain of the unknown. Sir Isaac Newton, one of the greatest scientists who ever lived, was one who did:

> I do not know what I may appear to the world; but to myself I seem to have been only like a boy playing on the seashore and diverting myself, now and then finding a smoother pebble or a prettier shell than ordinary, whilst the great ocean of truth lay all undiscovered before me.

More often philosophers sense the limitations of the present more quickly than do scientists, absorbed as the latter are in their immediate problems. The philosopher William James pointedly remarked upon the restrictive views of the "establishment" of his day (1895), particularly as manifested among his colleagues at Harvard:

> There is included in human nature an ingrained naturalism and materialism of mind which can only admit facts that are actually tangible. Of this sort of mind the entity called "Science" is the idol. Fondness for the word "scientist" is one of the notes by which you may know its votaries; and its short way of killing any opinion that it disbelieves in is to call it "unscientific." It must be granted that there is no slight excuse for this. Science has made such glorious leaps in the last 300 years . . . that it is no wonder if the worshippers of Science lose their heads. In this very University, accordingly, I have heard more than one teacher say that all the fundamental conceptions of truth have already been found by Science, and that the future has only the details of the picture to fill in. But the slightest reflection on the real conditions will suffice to show how barbaric such notions are. They show such a lack of scientific imagination that it is hard to see how one who is actively advancing any part of Science can make a statement so crude. Think how many absolutely new scientific conceptions have arisen in our generation, how many new problems have been formulated that were never thought of before, and then cast an eye upon the brevity of Science's career. Is this credible that such a mushroom knowledge, such a growth overnight at this, can represent more than the minutest glimpse of what the universe will really prove to be when adequately understood? No! Our Science is but a drop, our ignorance a sea. Whatever else be certain, this at least is certain: that the world of our present natural knowledge *is* enveloped in a larger world of some sort, of whose residual properties we at present can frame no positive idea.

Three-quarters of a century have passed since William James berated his Harvard colleagues; time has fully vindicated him. Though he could hardly have suspected it, the year 1895 was to be the first of "the thirty years that shook physics," that saw relativity, quantum mechanics, and many associated new concepts uproot the tenets of classical physics that were accepted by all physicists as the very rock foundation of the physical universe. The growth of our knowledge and technology has been exponential, yet we must say, unless we are both purblind and unutterably smug, that our ignorance is still a sea.

PART I

The UFO Phenomenon

INTRODUCTION: AN INNOCENT IN UFO LAND

After twenty-two years of "stewardship" of the UFO problem, the Air Force terminated its "Project Blue Book," the name given to the major portion of its UFO investigation program. Originally termed "Project Sign" and initiated in September, 1947, on February 11, 1949, it became "Project Grudge"; then from summer of 1951 to late 1960 it was called, "Project Blue Book." Code names are not supposed to have any special significance, but the reader may read into them whatever he wishes.

Throughout this period the project was located at the Wright-Patterson Air Force Base in Dayton, Ohio, first as part of the Air Technical Intelligence Center (ATIC) and later under the aegis of the Foreign Technology Division (FTD). The Air Force's formal public association with the UFO problem ended in December, 1969, when Secretary of the Air Force Robert C. Seamans officially terminated Project Blue Book, largely upon the recommendation of the Condon Report, the work of the Air Force-sponsored scientific group at the University of Colorado under the direction of Dr. E. U. Condon.

In my association with the UFO phenomenon I was somewhat like the proverbial innocent bystander who got shot. Project Sign needed an astronomer to weed out obvious cases of astronomical phenomena—meteors, planets, twinkling stars, and other natural occurrences that could give rise to the flying saucer reports then being received, and I was a natural choice. I was then director of Ohio State University's McMillin Observatory and, as such, the closest professional astronomer at hand.

Before I began my association with the Air Force, I had joined my scientific colleagues in many a hearty guffaw at the psychological postwar craze for flying saucers that seemed to be sweeping the country and at the naivete and gullibility of our fellow human beings who were being taken in by such obvious nonsense. It was thus almost in a sense of sport that I accepted the invitation to have a look at the flying saucer reports—they were called "flying saucers" then. I also had a feeling that I might be doing a service by helping to clear away nonscience. After all, wasn't this a golden opportunity to demonstrate to the public how the scientific method works, how the application of the impersonal and unbiased logic of the scientific method (I conveniently forgot my own bias for the moment) could be used to show that flying saucers were figments of the imagination? Although many of my colleagues at the university looked askance at my association with such "unscientific activity," I felt secure. I had ample files protection; as an astronomer I had been *invited* to examine the subject.

Such was my initiation and my inclination at the time. However, the opportunity to demonstrate to the public how the scientific method works, using the analysis of flying saucer reports as the vehicle, never materialized. While I was still working on my report for Project Sign, it became Project Grudge, and the Pentagon began to treat the subject with subtle ridicule. Furthermore, even though many UFO reports were not militarily classified, they were still by no means open to public examination. Such strictures effectively prevented letting the public in on the results of flying saucer investigations, let alone the process of investigation. The public was given only the end results—in cryptic news releases that, on the whole, left their questions unanswered and lowered the public's estimation of the Air Force's scientific image.

I played essentially no part in Project Grudge, and it was not until after the organization of Project Blue Book, under Captain Ruppelt in 1952, that I again became scientific consultant on UFO matters. Although my chief responsibility was as astronomical consultant, I concerned myself with all reports as they came in, each month reviewing current reports. Thus I became aware of some very interesting cases, most of which were submerged in a veritable quagmire of nonsense reports.

The termination of Project Blue Book heightened my sense of obligation to set forth my experiences, many of them startling, with the UFO problem and with the Air Force over a period of more than twenty years. Now I feel somewhat like a traveler returned from a long journey through unexplored, strange, and exotic lands, who finds it incumbent upon himself to set down an account of his travels and of the bizarre antics and customs of the "natives" of that strange land for the benefit of those who stayed at home.

The last twenty years have seen a plethora of books and articles on UFOs and flying saucers, but I have not contributed to that flood of literature except by submitting a few articles. I certainly do not wish to add just another book to the pile. I hope, rather, that the present work will be a positive contribution to the serious study of this subject. In any event, it is a view from within since I happened to be around when the Air Force needed an astronomer to help examine the rapidly accumulating pile of UFO reports. I have had an opportunity to read and study all the reports in the Blue Book files, to interview many hundreds of witnesses—the reporters of UFO experiences—and even to testify several times before Congressional groups which expressed considerable interest in the antics of the natives of UFO land.

I have often been asked whether I myself have had a UFO experience. The answer is no if I apply the tests I insist are necessary, which will be made clear in later chapters. On two separate occasions in the past twenty years I have seen an object and a light, respectively, that I could not readily explain, but since a possible, though not particularly probable, natural explanation exists, these two experiences do not fall within the definition of UFO used in this book. I have never experienced a close encounter (Chapter Four) and probably would not have reported it if I had, unless I had several reputable witnesses, but this does not surprise me. Statistics indicate that such sightings are indeed rare events, perhaps akin to the sighting of an extremely rare or unnamed species of bird (and how would you prove that on a walk through the mountains and woods you had sighted a California condor?) though not as rare as finding a coelacanth in the ocean depths. My experience with UFOs is secondhand, observed entirely through the eyes of others. The natives in UFO land are reports and the people who have made those reports. They are both worthy of discussion.

For the purpose of clarity I include a list of terms commonly used in the description of UFOs and in this text:

UFO Report—a statement by a person or persons judged responsible and psychologically normal by commonly accepted standards, describing a personal, visual or instrumentally aided perception of an object or light in the sky or on the ground and/or its assumed physical effects, that does not specify any known physical events, object, or process or any psychological event or process.

UFO Experience—the content of a UFO report.

UFO Phenomenon—the total class of the UFO Report and the UFO Experience.

UFOs—the existential correlates, if any, of the UFO Phenomenon; i.e., what if it exists, exists in its own right quite independently of the UFO Phenomenon.

The issue of existence is not amenable to a priori settlement but to settlement by investigation. If investigation indicates existence, this class may comprise:

(a) Hitherto undiscovered space-time items that conform to the laws of physics but require an extraordinary explanation.

(b) Hitherto undiscovered space-time items that conform to hitherto unformulated laws of physics.

(c) Hitherto undiscovered items, not in space, requiring nonphysical modes of explanation. If so, then these may be either unique products of individual or group mental action, conforming to known or unknown psychological laws, or something quite different from any of the above.

(New) Empirical Observations—any experience obtained directly through or with the aid of one or more human sense receptors that can be described in a report, which gives us information about what exists in its own right, quite apart from being thus experienced.

A New Empirical Observation is such an experience considered in relation to an existing body of information (e.g., scientific theory or theories) that is unable to incorporate it *without being revised or altered in fundamental respects.*

Flying Saucers—the original journalistic term for UFOs. In its long history, however, the term has been employed very broadly and with great confusion. To some it connotes a material craft capable of interstellar travel and of transporting intelligent extraterrestrial beings to earth. To some, on the other hand, it connotes *any* report of a seemingly unlikely sighting in the sky or on the ground, even when this is almost certainly due to a misperception of a normal object or event.

And to still others (generally members of flying saucer cults, or to groups of true believers), it signifies the visitation to earth of generally benign beings whose ostensible purpose is to communicate (generally to a relatively few selected and favored persons—almost invariably without witnesses) messages of "cosmic importance." These chosen recipients generally have repeated contact experiences, involving additional messages. The transmission of such messages to willing and uncritical true believers frequently leads, in turn, to the formation of a flying saucer cult, with the "communicator" or "contactee" the willing and obvious cult leader. Although relatively few in number, such flying saucer advocates have by their irrational acts strongly influenced public opinion—sometimes the opinions of learned persons such as Dr. Condon and some of his associates.

Clearly, flying saucers, whether defined as extraterrestrial craft, misperceptions, or highly mission-oriented carriers of cosmic knowledge to contactees, obviously do not satisfy the definition of UFOs since all of these definitions presuppose, a priori, the origin and nature of flying saucers.

CHAPTER ONE

The Laughter of Science

I know the moon and the stars, and I know shooting stars. I am not a young man. I have been born many years. I have been looking at the sky all my life. But I have never seen anything like this before. You are a white man. Can you tell me what it is?

—PAPUAN VILLAGE COUNSELOR

During an evening reception of several hundred astronomers at Victoria, British Columbia, in the summer of 1968, word spread that just outside the hall strangely maneuvering lights—UFOs—had been spotted. The news was met by casual banter and the giggling sound that often accompanies an embarrassing situation. Not one astronomer ventured outside in the summer night to see for himself.

Erwin Schrodinger, pioneer in quantum mechanics and a philosopher of science, wrote, "The first requirement of a scientist is that he be curious. He should be capable of being astonished and eager to find out."[1]

The scientific world has surely not been eager to find out about the UFO phenomenon and has expressed no inclination to astonishment. The almost universal attitude of scientists has been militantly negative. Indeed, it would seem that the reaction has been grossly out of proportion to the stimulus. The emotionally loaded, highly exaggerated reaction that has generally been exhibited by scientists to any mention of UFOs might be of considerable interest to psychologists.

Such reaction has been interesting to observe. I have attended many gatherings of scientists, both formal and informal, at which the subject of

UFOs has been brought up incidentally, either by chance or sometimes "innocently" by me in order to observe the reaction. I have found it amusing thus to set a cat among the pigeons, for the reaction has been out of keeping with the traditional "weigh and consider" stance of mature scientists. Frequently the reaction has been akin to that of a group of preteenagers watching a movie scene of exceptional tenderness or pathos quite beyond their years to appreciate: giggles and squirming suggest a defense against something the scientists cannot yet understand. It has seemed to me that such exhibitions by mature scientists are more than expressions of pity for the uninformed. Perhaps they are expressions of deepseated uncertainty or fear.

It is necessary here to distinguish two different classes of scientists who are confronted formally with the topic of UFOs: (1) those scientists who treat the UFO phenomenon with ridicule and contempt, refusing even to examine it, denouncing the subject out of hand; and (2) those scientists who maintain—or might come to believe after examination—that there is a strong possibility that UFOs are purely psychological phenomena, that is, generated wholly by individual or group mental activity. (No scientist who examines the subject objectively can claim for long that UFOs are solely the products of simple misidentification of normal objects and events.)

The views of the latter group are entitled to serious discussion and scientific debate, for the scientists have taken the trouble to examine the problem and accordingly should be heard. The views of the former group do not meet the conditions of scientific debate because there has been no examination of the data. Scientists of good standing have toured the country declaiming against the UFO phenomenon, refusing to answer questions from the floor while proudly pointing out that they haven't taken the trouble to examine "all the rubbish." The phenomenon of this modem witchhunt, the antithesis of what the scientific attitude stands for, is itself a phenomenon worthy of study. If all this UFO business is nonsense, why the overreaction on the part of established and highly respectable scientists? Is it a subconcious reaction to a challenge they are not prepared to accept?

Thomas Goudge, a noted Canadian philosopher of science, writes:

> "One of the most interesting facets of the UFO question to me is its bearing on the problems of how science advances. Roughly I would say that a necessary condition of scientific advancement is that allowance must be made for (1) genuinely new empirical observations and (2) new explanation schemes, including new basic concepts and new laws."[2]

Goudge points out that throughout history any successful explanation scheme, including twentieth-century physics, acts somewhat like an establishment and tends to resist admitting new empirical observations (unless they have been generated directly within the framework of that explanation scheme). Thus, for instance, most physical scientists were initially reluctant to admit now-accepted theories of meteorites, fossils, the circulation of the blood, bacteria, and, in our times, ball lightning, into the area of respectable science.

"For," Goudge continues, "if the establishment assimilates the new observations into the present explanation scheme, it implies that the empirical observations are not genuinely new. . . ." For example, scientists once were prepared to allow that meteorites existed not as stones from the sky but as stones that had been struck by lightning. This theory allowed assimilation of a new phenomenon into the accepted explanation scheme of the physical world about them. They could not admit that meteorites came from space. "Hence the present establishment view," Goudge concludes, "that UFO phenomena are either not really scientific data at all (or at any rate not data for physics) or else are nothing but misperceptions of familiar objects, events, etc. To take this approach is surely to reject a necessary condition of scientific advance."

The phrase "genuinely new empirical observations" is central to the entire UFO problem. Either UFO observations represent genuinely new empirical observations—that is, new in the sense that they do not fall immediately into place in the present scientific framework—or they simply are misperceptions and misidentifications. As far as UFOs are concerned, which is the case is not at all obvious except to those scientists who steadfastly refuse to dismiss the subject without consideration.

It is likely that many scientists would have given serious consideration and effort to the UFO problem had they been properly apprised of its content. Unfortunately, those few scientists who wished to be informed on the subject were forced to obtain information from the press, from sensational tabloid articles, and from pulp magazines generally catering to adventure, mystery, sex, and the sensational aspects of the occult. Until very recently no scientific journal carried any UFO information whatever, yet a recent bibliography of "UFO literature" of all sundry sorts ran to 400 pages. It would appear that the UFO became a problem for the librarian sooner than it did for the scientist.

Scientists are not the only group that is misinformed about the UFO dilemma. As the result of bad press, the public at large has accepted certain misconceptions about UFOs as true:

Only UFO buffs report UFO sightings. Oddly enough, almost exactly the opposite is true. The most coherent and articulate UFO reports come from people who have not given much thought to the subject and who generally are surprised and shocked by their experience. On the other hand, UFO buffs and believers of the cultist variety rarely make reports, and when they do, they are easily categorized by their incoherence.

This misconception was certainly in the mind of a most prominent scientist and an erstwhile colleague, Dr. Fred Whipple, director of the Smithsonian Astrophysical Observatory, for which I served for several years as associate director: "I will end with my now standard comment to newspaper reporters who ask me about UFOs. My reply is, 'I do not make public statements about the beliefs of religious cults.'"[3] (Faced with such a reaction, I made the proper answer: "Neither do I.")

UFOs are never reported by scientifically trained people. On the contrary, some of the very best reports have come from scientifically trained people. Unfortunately, such reports are rarely published in popular literature since these persons usually wish to avoid publicity and almost always request anonymity.

UFOs are reported by unreliable, unstable, and uneducated persons. Some reports are indeed generated by unreliable persons, who in daily life exaggerate other matters besides UFOs. But these people are the most apt to report misperceptions of common objects as UFOs. By the same token, however, these reporters are the most easily identified as such, and their reports are quickly eliminated from serious consideration. Only reports that remain puzzling to persons who by their training are capable of identifying the stimuli for the report (meteors, birds, balloons, etc.) are considered in this book as bona fide reports.

Reports are sometimes generated by uneducated people, but "uneducated" does not necessarily imply "unintelligent." Air crash investigators have found, for instance, that the best witnesses are teenaged boys, untrained but also unprejudiced in reporting.[4] In contrast, dullards rarely overcome their inherent inertia toward making written reports and frequently are incapable of composing an articulate report.

Very few reports are generated by mentally unstable persons. Psychiatrist Berthold Schwarz examined 3,400 mental patients without finding experiences related to UFOs.[5] His findings are supported by many colleagues, who found that there is an almost complete absence of UFO-related experiences among mental patients (they have, incidentally, little or no interest in the subject).

UFOs are synonymous with "little green men" and visitors from outer space. It is not known what UFOs are. To reject the phenomenon on the assumption that UFOs can arise from nothing except "space visitors" is to reject the phenomenon because one, for his own good reasons, rejects a theory of the origin of the phenomenon.

The chief objective of this book is to help to clear away these misconceptions by presenting *data* rather than by giving, *ex cathedra*, a pontifical pronouncement on the nature of UFOs. Before we examine the UFO experience further, it will help—indeed it is essential—to define as strictly as possible what the term UFO will mean throughout this book. It need not be a complex definition.

We can define the UFO simply as the reported perception of an object or light seen in the sky or upon the land the appearance, trajectory, and general dynamic and luminescent behavior of which do not suggest a logical, conventional explanation and which is not only mystifying to the original percipients but remains unidentified after close scrutiny of all available evidence by persons who are technically capable of making a common sense identification, if one is possible.

(For example, there are many thousands of people to whom the planet Venus is unknown, but UFO reports generated by this brilliant object in the evening or dawn sky will not fool an astronomer.)

Using this definition, I can say categorically that my own study over the past years has satisfied me on the following points:

(1) Reports of UFO observations that are valid for study exist quite apart from the pronouncements of "crackpots," religious fanatics, cultists, and UFO buffs.

A large number of initial UFO reports are readily identifiable by competent persons as misperceptions and misidentifications of known objects and events. Hence they must be deleted prior to any study aimed at determining whether any genuinely new empirical observations exist.

(2) A residue of UFO reports is not so identifiable. They may fall into one or more of the following categories:

(a) those that are global in distribution, coming from such widely separated locations as northern Canada, Australia, South America, Europe, and the United States;

(b) those made by competent, responsible, psychologically normal people—that is, by credible observers by all commonly accepted standards;

(c) those that contain descriptive terms that *collectively* do not specify any known physical event, object, or process and that do not specify any known psychological event or process;

(d) those that resist translation into terms that apply to known physical and/or psychological events, objects, processes, etc.

In the chapters that follow data to support these contentions are presented.

CHAPTER TWO

The UFO Experienced

The experience that I had on that June 8, 1966, morning will never be forgotten by me. Nothing since that sighting has convinced me that I was only thinking that I was seeing what I did see. I was upset for weeks after that experience; it scared the hell out of me. I was one of the combat crew members that sighted the first German jet fighter flights in World War II. The Air Force tried to convince us that we were seeing things then also.

—FROM A PERSONAL LETTER TO THE AUTHOR

In my years of experience in the interrogation of UFO reporters, one fact stands out: invariably I have had the feeling that I was talking to someone who was describing a *very real* event. To them it represented an outstanding experience, vivid and not at all dreamlike, an event for which the observer was usually totally unprepared—something soon recognized as being beyond comprehension. To the reporter and to any companions who shared the experience the event remained unexplained and the phenomenon unidentified even after serious attempts at a logical explanation had been made. The experience had the reality of a tangible physical event, on a par with, for example, the perception of an automobile accident or of an elephant performing in a circus, *except for one thing*: whereas reporters have an adequate vocabulary to describe automobiles and elephants, they are almost always at an embarrassing loss for words to describe their UFO experience.

In my experience in interrogating witnesses one phrase has been repeated over and over again: "I never saw anything like this in my life." But

I have also found that the reporters of the UFO experience try their best to describe and explain their experience in conventional terms. They almost always attempt to find—even force upon the lack of fact, if necessary—a natural explanation. In direct contradiction to what we are often told, that people see what they wish to see, my work with UFO reporters of high caliber indicates that they wish to see or to explain their observations in terms of the familiar. A typical statement is: "At first I thought it might be an accident up ahead on the road—the lights looked something like flasher beacons on squad cars. Then I realized that the lights were too high, and then I thought maybe it was an airplane in trouble coming in for a crash landing with power off, since I didn't hear any sound. Then I realized it was no aircraft."

I have seen this process of going from the simple, quick description and explanation, step by step, to the realization that no conventional description would suffice (escalation of hypotheses) happen far too often to be able to subscribe to the idea that the UFO reporter has, for inner psychic reasons, unconscious images, or desires, used a simple, normal stimulus as a vehicle for the expression of deep-seated inner needs. The experience is for the reporter unique and intensely baffling; there is an unbridgeable gap between the experience and being able to fit it into a rational description and explanation.

It is indeed difficult to dismiss, out of hand, experiences that lead a person of obvious substance to say, in all sincerity:

> I only know that I have never seen anything in the sky shaped quite like it, nor have I ever seen any plane which moves at such a great speed.[1]

> It was just like looking up under an airplane, just as if an airplane were standing there. Just perfectly motionless and no noise whatever. We watched this possibly for five minutes—then the thing got a tremendous burst of speed and sped right off. No sound whatsoever.[2]

> The RCMP (Royal Canadian Mounted Police) asked me at the time if I thought it was a helicopter above the clouds, with this object dangling on a rope. Now that's the silliest explanation I've ever heard.[3]

These are by no means exceptional quotations. Dozens of others, gathered from my own as well as from Project Blue Book files, could fill this chapter, and much more. And many of them concern experiences shared by more than one participant. Still, the words alone do not convey the human experience described by the observer. Many times I have mused, "How is it possible that this apparently sane, steady, responsible person is standing there telling me this story with all seeming sincerity? Can they possibly be acting this out? Could they be such a good actor?' And if so, to what end? They surely must know that this incredible tale could set them up as a target for merciless ridicule." Here are two other reactions:

> I heard the dog barking outside. It was not a normal barking, so I finally got a little angry with him and went outside. I noticed the horses were quire skittish and were running around the pasture. I looked up to see what the horses were worried about. I saw this object sitting up in the air—it would be about 400 to 500 feet off the ground. I asked my friend to come out and have a look to see what I saw or if I was going off my rocker. That person came out, took one look, screeched, and ran back into the house . . .
>
> I assumed as a matter of course that it was a totally new invention and fervently hoped that the inventors were our own people, for this was still prior to VJ-Day. I made up my mind that I would tell no one of my sighting until the news became public.

Sometimes the reports or interviews contain frank and artless remarks, which nonetheless attest to the "realness" of the event for the witness. This comment came from four boys at Woodbury Forest School about a sighting on February 15, 1967: "This is the truth, and there is no hoax implied since that is a serious offense at this school."

From three Boy Scouts in Richardson, Texas: "Mike, Craig, and I are Boy Scouts in Troop 73 . . . and we give you our Scout's Honor that this is not a hoax or optical illusion."

It would be hard to beat the following remark for ingenuousness: "What I am trying to say is that I didn't use any trick photography because I don't know how yet . . ." This statement was made in a report of a sighting in New Jersey on December 26, 1967.

Finally we have this plaintive appeal (from a letter to Blue Book describing a sighting of a cigar-shaped object on January 19, 1967): "Although I am only a child, please believe me."

It is often the peripheral remarks of a mature and serious nature that emphasize the vividness of the reporter's experience. This comment was made by a Trans-Australia Airlines pilot with some 11,500 hours of flying experience: "I had always scoffed at these reports, but I saw it. We all saw it. It was under intelligent control, and it was certainly no known aircraft."[4]

The following is a statement from a man who flew fifty combat missions in World War II. He is a holder of five Air Medals and twelve Bronze Major Battle Stars, and he is, presumably, not easily alarmed: "There was no sound, and it was as long as a commercial airliner but had no markings . . . My body reacted as if I had just experienced a 'close shave' with danger. For the remainder of the day I was somewhat emotionally upset."[5]

The objects, or apparitions, being described are discussed in some detail later. Here I wish simply to convey to the reader as best I can the fact that the UFO experience is to the reporter an extremely real event.

Often I wondered as I listened to a graphic account of a UFO experience, "But why are they telling me this?" I realized at length that the reporters were telling me because they wanted me to *explain* their experiences to them. They had been profoundly affected, and they wanted an explanation that would comfortably fit into their world picture so that they could be relieved of the burden of the frightening unknown. Their disappointment was genuine when I was forced to tell them that I knew little more about it than they did. I knew only that their experience was not unique, that it had been recounted in many parts of the world.

Though it cannot be explained—yet—the UFO experience (as UFO is defined in this book) has every semblance of being a real event to the UFO reporter and his companions. That is our starting point.

CHAPTER THREE

The UFO Reported

The unquestioned reliability of the observer, together with the clear visibility existing at the time of the sighting, indicate that the objects were observed. The probable cause of such sightings opens itself only to conjecture and leaves no logical explanation based on the facts at hand.

—FROM AN OFFICIAL INVESTIGATION REPORT
MADE BY AN AIR FORCE CAPTAIN

What kind of person has a UFO experience? Are they representative of a cross section of the populace or are they something "special"? In trying to answer such questions, we immediately face two conditions. First, we can study only those who *report* having had a UFO experience. There is much evidence that relatively few who have such an experience report it. Second, we consequently cannot ask what *kind* of person has a UFO experience but only what kind of person *reports* that they have seen a UFO.*

What sort of person fills out a long questionnaire about such a sighting or writes an articulate account of it in the face of almost certain ridicule? Are they charlatans, pixies, psychotic, or responsible citizens who feel it is their duty to make a report? The only type of reporter the serious student

* For that reason it is better to speak of a UFO *reporter* rather than of a UFO *observer* since if it should prove that UFOs are not *real*, there could be no UFO *observers*, but there could be, and indeed are, UFO *reporters*.

needs—and should—bother with is the sort of person who wrote the following letter:

> "I am postmaster here at — and I hesitated before reporting this subject to the postal inspector. But after a great deal of serious thinking, I felt I ould not be a good Amerian citizen if I did not ask these questions, 'What was the lighted object and where did it come from?'"[1]

The reliable UFO reporter is generally acknowledged in his community to be a stable, reputable person, accustomed to responsibility—with a good job and known to be honest in their dealings with others.

It has been my experience that UFO reporters have little in common by way of background. They come from all walks of life. Yet in addition to a shared reputation for probity they often experience a marked reticence to talk about their experiences, at least until they are assured of the interrogator's sincerity and seriousness.

> What I have written . . . is for you and your research work . . . I have never reported any of this. But I do believe you should have this information in detail. But for no newspaper, no reporters . . . I am still reluctant, but somehow I feel you are the right person.[2]

> I have discussed this matter only with two men—one a prominent manager in our area, and the other my pastor.[3]

> I can tell you one thing—if I ever see another one, mum's the word. We called the city police first to ask if anybody had reported a UFO, and the man at the telephone laughed so long and loud that I'm sure he must have almost fell off his chair . . . [the paper] ran some darn smart aleck story that made all of us look like fools.[4]

Such expressions of embarrassment and hesitation are frequently encountered, and the very fact that the reporters, in the face of almost certain scorn, have persisted in making a report indicates a genuine feeling that the information is of importance and should be transmitted to someone. Reporters' actions likewise indicate a haunting curiosity about their experiences, a feeling frequently so great that it alone is enough to make the reporters brave the almost certain ridicule.

Why this emphasis on the *character* of the reporter? Given the fact that in most other areas of science, electronic and optical instruments supply us with the data for analysis, the nature of the UFO reporter is of paramount importance. In this area of scientific inquiry the UFO reporter is our *only* datagathering instrument.

In science it is standard practice to calibrate one's instruments. No astronomer, for instance, would accept measures of the velocities of distant galaxies obtained by means of an uncalibrated spectrograph. However, if such an instrument had given consistently good results in the past, had frequently been tested, and had not recently experienced any recent jarring shocks, the astronomer will usually accept its results without further checking.

The parallel for us is, of course, obvious: if our UFO reporter has by past actions and performance shown a high degree of reliability and responsibility and is known to be stable and not "out of adjustment," then we have no a priori reason to distrust their coherent report, particularly when it is made in concert with several other "human instruments" also of acceptable reliability.

While a battery of tests designed to determine the veracity and stability of a person is available today, because of the scientific establishment's refusal to take the matter seriously, the tests are not usually readily available to the UFO investigator, even though the UFO reporter frequently is willing to undergo such tests (an important point of fact in itself). We must, therefore, usually content ourselves with judging—from the person's vocation, family life, the manner in which they discharge responsibilities and comport themselves—their "credibility index." We must decide whether the composite credibility index that can be associated with a report when several persons contributed to that report makes the material worth consideration.

Essentially, the crucial question is, did what the reporters say happened *really happen*? We may equally well ask: if, when a speedometer indicates a speed of 80 miles per hour, is an automobile really going 80 miles per hour? Is the speedometer to be trusted? Or, are the reporters to be trusted? Obviously the human mind cannot be equated with a speedometer. There are too many stories of people who have led exemplary lives and have suddenly gone berserk, committed a murder or a robbery or exhibited some other act of outrageous antisocial behavior. Still, it is most unlikely that *several* persons would *simultaneously* "break" and commit such an act entirely out of keeping with their characters—or jointly commit the "crime" of reporting a UFO. And provided we do not put too much weight on any one single report, there is no reason not, at least at first, to believe them.

"Why *shouldn't* we believe what several UFO reporters of established personal reputation tell us?" is just as valid a question as "Why should we believe them?" Criteria for disbelief and for belief are on a par. For example, what a priori reason do we have *not* to believe the following direct statements from, according to all other evidence, reputable people:

> I have traveled US Highway 285 over Kenosha Pass for over twenty years, day and night; This was my first sighting of a UFO.[5]

> We own a business in our home town, and we are well known, so I am not the sort of person that would make a crank call. I don't know what it was that we saw, but we saw something, and it was as real as real can be.[6]

> Before you throw this away as just another crank letter, consider that I am a fifty-one-year-old mathematics teacher who has never suffered from mental illness nor been convicted of a crime. I have never knowingly had hallucinations nor been described as neurotic . . . nor do I seek publicity. Quite the opposite is the case, for it has been my experience that anyone who claims to have witnessed a genuine UFO is regarded as some kind of nut. Yet I have unquestionably and clearly sighted an as-yet-unexplained flying object.[7]

(These are just a small sample of the types of statements I have listened to, read in personal letters, and found in official UFO reports.)

It is interesting to note, as substantiation of the theory of the credibility of reliable witnesses, that in those instances in which fake UFOs have been deliberately contrived to test public reaction—hot air balloons and flares dropped from airplanes are examples—the resulting UFO reports were not only invariably far fewer than the experimenter expected but of interest more for what they did *not* report than for what they did. Occasionally a fanciful UFO report is generated as a result of such a test, but it fails to meet the test of acceptance because it does not square with what others have reported about the same event—often solely because of its internal inconsistency and incoherence.

The almost complete absence in such reports of occupants, interference with automobile ignition systems, landing marks, and other physical effects on the ground, and the many other things characteristic of reports of Close Encounters is eminently noteworthy. Comparison of accounts from various reporters adds up to a perfectly clear picture of the actual event—a hot air balloon, a flare, or a scientific experiment. The duration of the event, the direction of motion of the balloons or flares, and even the colors are reasonably well-described.

There are exaggerations, of course, and considerable latitude in descriptions (but hardly greater than one gets in collective accounts of fires, automobile accidents, etc.); but one is rarely left in no doubt about what actually happened. Descriptions of fires or airplane crashes made by seemingly reputable witnesses may vary greatly in detail, but one is never in doubt that a fire or a plane crash and not a bank robbery is being described. One does not get collective statements from several witnesses to a hot air balloon UFO that they saw a UFO with portholes, antennas, occupants, traveling against the wind, changing direction abruptly, and finally taking off at a 45-degree angle with high speed. One is quickly led by the study of such reports to the actual event that caused them.

True, occasionally a lone witness of low credibility will make a highly imaginative report, generated by an obviously natural event. But such reports are a warning to beware of UFO reports from single witnesses; one can never be too careful, even in instances in which the reporter is adjudged to be reliable.

For all these reasons, then, there are no a priori reasons for dismissing such statements out of hand. The crux of the UFO reporter problem is simply that perfectly incredible accounts of events are given by seemingly credible persons—often by several such persons. Of course, what the UFO reporter says *really* happened is so difficult to accept, so very difficult a pill to swallow, that any scientist who has not deeply studied the UFO problem will, by the very nature of his training and temperament, be almost irresistibly inclined to reject the testimony of the witnesses outright. Not to do so would be to reject his faith in his rational universe. Yet not to do so also involves the rejection of material that will not "just go away" if it is ignored.

Responsible persons *have reported* phenomena that defy scientific explanation, and until *unimpeachable* radar records and photographic evidence are at hand, the UFO reporter, who is all we have to depend on, must be heard out. There are just too many of them, from all parts of the world, to disregard their word. To do so would be scientific bigotry, and we must not stand accused of such a charge.

CHAPTER FOUR

On the Strangeness of UFO Reports

I should add that I have never been a believer in UFOs before, but this one was so unexplainable by our present standards that it has me wondering.

—FROM A LETTER TO DR. CONDON REPORTING A UFO

UFOs exist, for most of us, as *reports*, and most of us consider such reports sensationalized stories in pulp magazines and as scattered news items: "Police Track Mystery Object" or "Aircraft Buzzed by Glowing UFO." Such newspaper accounts at one time became so commonplace that editors ceased to consider them newsworthy. To the UFO percipient, in strong contrast, the UFO exists as an intensely *personal* experience. The gap between the two approaches a yawning chasm.

The problem is compounded by the fact that most UFO reports are frustrating in the extreme. They contain so few facts! This lack alone has deterred several scientists from devoting time to the matter, for they expect to find data they wish to study in the form to which they are accustomed: instrument readings, photographs, charts, graphs, and tables, with as much of the data as possible in quantitative numerical form.

Yet the paucity of data is more often the fault of the investigator than of the original reporter. The latter comes upon the experience suddenly, totally unprepared. They generally are so shocked and surprised that careful sequential observation and reporting are impossible. The skillful interrogator can, of course, extract details from the reporters that they had noticed only incompletely or had believed to be irrelevant. Most people faced with witnessing a sudden and shocking automobile accident do not go about methodically making measurements, checking times, durations, length of skid marks, condition of the weather, and other related evidence. However, in retelling the incident to a competent interrogator the latter can deduce and extract through calm and adroit questioning a surprising amount of information from the witness.

By contrast, very frequently Air Force investigators, imbued with the official philosophy that UFOs are delusions, make only a perfunctory interrogation (why spend time on something that is meaningless in the first place?).

Still, there exist UFO reports that are coherent, sequential narrative accounts of these strange human experiences. Largely because there has been no mechanism for bringing these reports to general attention, they seem to be far too strange to be believed. They don't fit the established *conceptual framework* of modern physical science. It is about as difficult to put oneself into a "belief framework" and accept a host of UFO reports as having described actual events as, for example, it would have been for Newton to have accepted the basic concepts of quantum mechanics.

Yet the strangeness of UFO reports does fall into fairly definite patterns. The "strangeness-spread" of UFO reports is quite limited. We do not, for instance, receive reports of dinosaurs seen flying upside down, Unidentified Sailing Objects, or strange objects that burrow into the ground.

A critic of the UFO scene once remarked, "unexplained sightings do not constitute evidence in favor of flying saucers any more than they constitute evidence in favor of flying pink elephants." What he failed to realize was that the strangeness spectrum of UFO reports is so narrow that not only have flying pink elephants never been reported but also a *definite pattern* of strange craft has. If UFOs indeed are figments of the imagination, it is strange that the imaginations of those who report UFOs from over the world should be so restricted.

Precisely because the spectrum of reports of strange sightings is narrow can they be studied. If each strange report was unique and their totality ran the gamut of all conceivable strange accounts, scientific investigation of such a chaotic panoply would be impossible. Scientific study presupposes data patterns and a measure of repeatability, and by and large, UFO reports lend themselves to classification within their domains of strangeness. It is these we shall pursue.

Turning, then, to the *content* of UFO reports, let us assume that we have eliminated all those reports which do not fit the definition of UFO as used earlier; that is, the dross from the original mass of "raw" reports—all reports that *can* be explained with good reason as balloons, aircraft, meteors, etc. (Such reports represent the "garbage" in the problem. If we incorporate that in our studies, the computer-age adage, Garbage in, garbage out, will surely apply. This has been the trap that UFO investigations in the past have not been able to avoid.*)

In terms of scientific study, the only significant UFO reports are, as we have seen, UFO reports that remain puzzling *after* competent investigation has been conducted. Only these can be termed reports of UFOs. The stimulus for these reports is truly unknown—that is, the reporters have passed a reliability screening, and the known possible stimuli have passed a physical explanation screening. Thousands of such reports exist; there are about 700 acknowledged cases in Blue Book files alone, and many others are contained in the files of UFO organizations and private investigators.

Each such screened report demands an answer to two distinct questions: What does it *say* happened? What is the *probability* that it happened? We

* Many critics maintain that all UFO reports are garbage. Since a large portion of the original, unfiltered reports are clearly the result of misperception, critics say that investigation in depth would reveal that the entire body of UFO phenomena can be so characterized. Such arguments assume that all UFO reports belong to the same statistical population and that the deviants, the truly interesting UFO reports, are merely extremes in that population. One might with equal justice say while plotting the variation in sizes of oranges that watermelons are merely "the tail end of the distribution curve" of the sizes of oranges.

can make those two questions the basis of a very helpful two-dimensional arrangement of UFO reports. Each report that has satisfied the definition of UFO used in this book can be assigned two numbers: its *Strangeness Rating* and its *Probability Rating*.

The Strangeness Rating is, to express it loosely, a measure of how "oddball" a report is within its particular broad classification. More precisely, it can be taken as a measure of the number of information bits the report contains, each of which is difficult to explain in common-sense terms. A light seen in the night sky the trajectory of which cannot be ascribed to a balloon, aircraft, etc., would nevertheless have a low Strangeness Rating because there is only one strange thing about the report to explain: its motion. A report of a weird craft that descended to within 100 feet of a car on a lonely road, caused the car's engine to die, its radio to stop, and its lights to go out, left marks on the nearby ground, and appeared to be under intelligent control receives a high Strangeness Rating because it contains a number of separate very strange items, each of which outrages common sense.

As we have seen, in the absence of hard-core evidence in the form of movies, detailed close-up shots, and so forth, we must depend greatly on the credibility of the principal reporter and his witnesses. Clearly, a report made by several independent persons, each of obvious sanity and solid general reputation, deserves more serious attention as probably having actually happened than a report made by a lone person with a none too savory record for veracity in past dealings with his fellow man.

This still leaves open a wide range of probability as to whether the strange event occurred as stated. Several judgment factors enter here as to whether what these otherwise reputable people reported on a particular occasion can be accepted—and with what probability. How much would one bet, even considering the qualifications of the reporters, that what was reported really happened as reported?*

---

* The philosopher Hume proposed a betting criterion as a way of measuring strength of belief. We can hardly do better.

Assessment of the Probability Rating of a report becomes a highly subjective matter. We start with the assessed credibility of the individuals concerned in the report, and we estimate to what degree, given the circumstances at this particular time, the reporters could have erred. Factors that must be considered here are internal consistency of the given report, consistency among several reports of the same incident, the manner in which the report was made, the conviction transmitted by the reporter to the interrogator, and finally, that subtle judgment of "how it all hangs together." It would be most helpful in the Probability Rating assignment if lie detector and other psychological tests were available. Likewise, a doctor's statement on the state of the reporter's health at the time or information of any severe emotional disturbance just prior to the time of the reported event would be helpful. Ideally, a meaningful Probability Rating would require the judgment of more than one person.

Such luxury of input is rarely available. One must make do with the material and facilities at hand. In my own work I have found it relatively easy to assign the Strangeness number (I use from 1 to 10) but difficult to assign a Probability Rating. Certainty (P = 10) is, of course, not practically attainable; P = 0 is likewise impossible under the circumstances since the original report would not have been admitted for consideration. The number of persons involved in the report, especially if individual reports are made, is most helpful. I do not assign a Probability Rating greater than 3 to any report coming from a single reporter, and then only when it is established that they have a very solid reputation. This is not to denigrate the individual but merely to safeguard against the possibility that the single meritorious reporter might have been honestly mistaken about what was experienced.

When the report "hangs together" and I honestly cannot find reason to doubt the word of the reporters—that is, unless I deliberately and with no reason choose to call them all liars—I assign a Probability of 5 or greater. Assignments to the upper right-hand region of the diagram showing the S-P indexes for the cases considered in this book (the symbols used for case classifications are explained below) are sparse because of the severity of the criteria employed. In fact, however, I discovered that a report accorded an S of 3 and a P of 5 (or a combined index of SP = 35) in every respect should command attention and challenge science.

The symbols used in the S-P diagram refer to the classification of the content of the report itself, independently of the reporter. The classification system itself is an empirical one, based on the reported manner of the UFO observation. It presupposes no theory of origin of UFOs but is helpful in delineating the most prevalent patterns found in UFO reports.

The classification has two main divisions: (I) those reports in which the UFO is described as having been observed at some distance; (II) those involving close-range sightings. The dividing line is not very sharp, but Close Encounter cases are those in which the objects were sighted at sufficiently close range (generally less than 500 feet) to be seen as extended areas rather than as near-points and so that considerable detail could be noted about them. The Close Encounter cases in category II clearly are apt to yield more strangeness information bits than the cases in category I since the witnesses presumably would have opportunity to observe colors, protrusions, sounds, dimensions, structural details, linear and rotational motion, "occupants," and any interaction of the UFO with the environment. The more distant UFOs will almost always have a lower S rating simply because there was not as much to observe and hence to explain.

The more distant UFOs I have arbitrarily divided into three categories: (1) those seen at night, which we will call Nocturnal Lights (designated N in the diagram); (2) those seen in the daytime, which we will call Daylight Discs (designated by D in the diagram), so called because the prevalent shape reported is oval or disc-like, although it should be understood that the term is rather loosely applied; and (3) Radar-Visual, those reported through the medium of radar (designated R in the diagram). In my own work I have chosen to exclude UFO observations made by radar alone because of a lack of a suitable filter to establish beyond reasonable doubt that the radar observation cannot be explained by natural causes (malfunctions, anomalous propagation, extraordinary meteorological conditions, flocks of "invisible" birds, swarms of high-flying insects, and so forth).

9		NNN ND	N	D	I	I	RI	RRI		9
8	NN	NNN ND	NNNN NNDD	RIIIN NND	RRRR IIID	RRI ID	IIII III	II II II II II III	III III	8
7	NN N	NNNN DD	RNNN NNNN DDDD	NDD DD	RRRI IIIIIN DDD	RIID	IIIIII II II II II II D	II II II II II	III III	7
6	NN	NNND	NDDD DDDD	DD	RI		II II II II II	II II II II II II	II III III	6
5	NN D	NNDD DDDD	IDD DD	RIID DD	IIIIII DDD	IIII	IIII II II II	I II II II II II II II II II	II	5
4	NN	NNNN DDDD	D	DD	I		I	II II III	III	4
3	N	ND	INN							3
2	DD	N								2
1										1
	1	2	3	4	5	6	7	8	9	

P = PROBABILITY OF EVENT AS DESCRIBED
(Credibility; Reliability)

S = STRANGENESS INDEX
(Measure of Items Requiring Explanation)

S–P DIAGRAM

When radar experts disagree among themselves as to the causes of "bogies" or "angels," I feel it is wisest to avoid introducing such evidence. When, however, visual observations *accompany* the radar observation, and if it can be established that the two types of observation refer with high probability to the same event, the radar observations become a powerful adjunct to the visual observation. In this book I use only such Radar-Visual cases; some of the very best UFO reports fall in this category.

The Nocturnal Lights and Daylight Discs may not be mutually exclusive, but at night almost invariably only the brightness, color, and motion of a light are reported. Rarely is the object noted to which the light is presumably attached (this is purely an assumption; the UFO may be nothing more than the light). Nocturnal Lights form a sizeable group of the true UFO reports.*

* Of course, before the screening process is undergone, reports of night lights constitute the great majority of the input. Bright planets, satellites, meteors and special aircraft missions are the preponderance.

The second major division of UFO reports comprises the Close Encounter cases. Here also there appear to be three natural subdivisions, which we can call, respectively, *Close Encounters of the First, Second, and Third Kinds* (designated in the diagram by the numbers I, II, and III, respectively).

Close Encounters of the First Kind: this category is the simple Close Encounter, in which the reported UFO is seen at close range but there is no interaction with the environment (other than trauma on the part of the observer).

Close Encounters of the Second Kind: these are similar to the First Kind except that physical effects on both animate and inanimate material are noted. Vegetation is often reported as having been pressed down, burned, or scorched; tree branches are reported broken; animals are frightened, sometimes to the extent of physically injuring themselves in their fright. Inanimate objects, most often vehicles, are reported as becoming momentarily disabled, their engines killed, radios stopped, and headlights dimmed or extinguished. In such cases the vehicles reportedly return to normal operation after the UFO has left the scene.

Close Encounters of the Third Kind: in these cases the presence of occupants in or about the UFO is reported. Here a sharp distinction must be made between cases involving reports of the presence of presumably intelligent beings in the spacecraft and the so-called contactee cases.

In general, the latter reports are stopped at the gate by the screening process. The reader will recall that implicit in our definition of UFO is the basic credibility of the reporter (unexplained reports made by ostensibly sensible, rational, and reputable persons). The contactee cases are characterized by a favored human intermediary, an almost always solitary contact who somehow has the special attribute of being able to see UFOs and to communicate with their crew almost at will (often by mental telepathy). Such persons not only frequently turn out to be pseudoreligious fanatics but also invariably have a low credibility value, bringing us regular messages from the space travelers with singularly little content. The messages are usually addressed to all of humanity to "be good, stop fighting, live in love and brotherhood, ban the bomb, stop polluting the atmosphere" and other worthy platitudes. The contactees often regard themselves as messianically charged to deliver the message on a broad basis; hence several flying saucer cults have from time to time sprung up. They regard themselves definitely as having been "chosen" and utterly disregard (if, indeed, they were capable of grasping it) the statistical improbability that one person, on a random basis, should be able to have many repeated UFO experiences (often on a nearly weekly basis), while the majority of humanity lives out a lifetime without having even one UFO experience. The "repeater" aspect of some UFO reporters is sufficient cause, in my opinion, to exclude their reports from further consideration, at least in the present study.*

I must emphasize that contactee reports are *not* classed as Close Encounters of the Third Kind. It is unfortunate, to say the least, that reports such as these have brought down upon the entire UFO problem the opprobrium and ridicule of scientists and public alike, keeping alive the popular image of "little green men" and the fictional atmosphere surrounding that aspect of the subject.

The typical Close Encounter of the Third Kind happens to the *same* sorts of persons who experience all other types of UFOs, representing the same cross section of the public. The experience comes upon these reporters just as unexpectedly and surprises them just as much as it does the reporters of the other types of Close Encounters. These reporters are in no way special. They are not religious fanatics; they are more apt to be police officers, business men and women, schoolteachers, and other respectable citizens. Almost invariably their UFO involvement is a one-time experience (whereas as we have seen, the contactee cases almost always involve rampant repeaters), and the sighting of occupants is generally a peripheral matter. The occupants in these cases almost never make an attempt to communicate; in contrast, they invariably are reported to scamper away or back into their craft and fly out of sight. They do not seem to have any messages for mankind—except "Don't bother me."

* Of course, perhaps the possibility that there are indeed "chosen ones" deliberately picked by UFO occupants for a special mission should not be completely disregarded. In that event, however, one is reminded of the Englishman's quip: "How unfortunate for these space visitors—every time to have picked a 'kook'!"

We thus have six categories of UFO reports, three in each broad division, to discuss. The classification is based solely on the manner in which the UFOs were reported to have been observed. The categories are obviously not mutually exclusive; a Daylight Disc seen from close by would become a Close Encounter; a Nocturnal Light seen by daylight might well be a Daylight Disc; and so on. It is convenient to discuss UFO reports in these categories simply because the data to be described are closely dependent on the manner in which they were experienced. If all reports in each separate category are discussed together, the patterns inherent in each are most directly delineated.

Finally, it should be remarked that when, in the original screening process, it is determined that the stimulus for the UFO report was indeed a natural event or object, the report does not generally fall easily into any of the six described categories. A UFO report generated by a hot air balloon does not contain the most often repeated feature of the typical Nocturnal Light. An aircraft fuselage glistening in the sun, reported by some untutored person as a UFO, is not reported to rush away at incredible speeds. Flares dropped from airplanes (which have often given rise to UFO reports) are not reported as having stopped cars, frightened animals, or cavorted about the sky; nor do the reports contain reference to occupants or to oval-shaped craft hovering six feet off the ground.

Having now briefly examined the nature of the UFO experience and the persons who report such experiences, and having classified the UFO reports into six convenient categories and established a system for the rating of UFO reports, let us now turn to the core of the book, the data available for study. Then, with this in mind, we shall proceed to a survey of how these data have been treated in the past, first by the Air Force and, more recently, by the Condon committee. Finally we will arrive at my suggestions for a positive program for the study of the UFO phenomenon.

PART II

The Data and the Problems

INTRODUCTION: THE PROTOTYPES

The problem central to this treatise is whether there exist, in the considerable body of data on reported UFOs, any genuinely new empirical observations calling for new explanation schemes. Very little ought to—or could—be said about what those new explanation schemes might be before a thorough examination of the data has been undertaken; this would be truly putting the cart before the horse. In such a controversial subject, which so frequently has triggered highly emotional reactions, examination of the data must come first; only then may we arrive at any judgment about new empirical observations. Indulging in explanation schemes before we know what there is to be explained is an armchair luxury.

One might be tempted to be less rigid on this point were the data of the hard-core variety, the kind with which physical scientists are accustomed to dealing in laboratory experiments. But from the standpoint of the scientist, the data in this problem are most unsatisfactory. They are more apt to be anecdotal than quantitative, more akin to tales told by the fireside than to instrument readings, and not verifiable by repeating the experiment.

The facts are not strictly scientific. Yet the data nonetheless form a fascinating and provocative field of study for those whose temperaments are not outraged by the character of the information. And it should be remembered that there are those whose fields of study abound with equally unsatisfactory data. Anthropologists, psychologists, and even meteorologists deal daily with evidential and circumstantial data that must be fitted together like pieces of a jigsaw puzzle. Lawyers and judges must weigh and consider conflicting evidence; military intelligence agents occasionally attempt to fashion a whole picture out of extremely fragmentary bits. Indeed, what constitutes hard-core data for one field of study may not be considered so for another. We may, therefore, examine the UFO data without reference to whether it meets the hard-core requirements of any particular field. Rather, we will examine, as objectively as possible, a specially selected series of data: accounts that were made, in each instance, by at least two persons of demonstrated mental competence and sense of responsibility, accounts that do not yield solutions except by the trivial and self-defeating artifice of rejection out of hand.

To this end, we may construct a paradigm for each of the observational categories delineated in the last chapter, drawing, for these prototypes, upon examples in whole or in part from cases I have personally studied. These archetypes will serve us better than would a review, perforce a brief one, of a whole series of individual cases.*

In my own work with this phenomenon I maintain three separate files for variously collected material: one contains highly selected cases with responsible observers, another has cases that might have been eligible for the selected files but for which there is not sufficient information about the observers to determine their reliability, and the third, a catch-all file, teems with reports that are scarcely above the caliber of a brief newspaper report, with many pertinent data missing and little or nothing said about the witnesses. Even the latter cases form a pattern and would probably be useful in statistical studies, though they are virtually useless for detailed studies.

All three files have about the same frequency of occurrence according to dates of the reported events; generally when newspaper accounts abound, so also do well-documented reports from responsible observers. There is nothing in the evidence to support the claim that the well-documented reports are spawned by a wave of loosely reported, sketchy accounts in the

* I personally have found it extremely difficult to deal with what essentially is a catalog of one UFO case after another, each briefly described but with the details and documentation omitted. The mind boggles at the repetitive strangeness and finds it difficult to digest and to order, in any logical manner, the veritable feast of strange accounts.

press. Rather, it might be argued that the former are simply the relatively few well-documented instances that might be expected to be found when there is a general period of UFO activity.

Since it has been my obligation over the years, as consultant to the Air Force, to try to separate the "signal" from the "noise," to wade through and judge the mass of vague and incomplete data, we can benefit from that experience and can short circuit much tribulation by examining what the accounts in each category essentially have *in common.* To that end, in the following chapters some dozen examples in each category have been chosen. The quality of the reporters involved in the cases has been evaluated, and the essential features that characterize that category have been set forth. References to the actual cases used are given in Appendix 1.

As part of the evaluation of the reporters, it is of interest to include many of the spontaneous reactions of the reporters to the event. Such instantaneous and ingenuous personal remarks and reactions help to characterize the reporters and to illuminate the extraordinary event. In the last analysis, the reporters or witnesses must take the center of our stage; they are our actors, and unless we know all we can about them, we might find to our embarrassment that we have "a tale told by an idiot . . . signifying nothing."

The cases in the six categories for which we seek prototypes have, of course, been passed through the filters described in Chapters 3 and 4, and for each of the more than sixty UFO reports used in the next several chapters I have not been able to find a logical commonplace explanation—unless, that is, I assume that the more than 250 reporters were, in truth, idiots.

CHAPTER FIVE

Nocturnal Lights

They [lights] appeared beneath the clouds, their color a rather bright red. As they approached the ship they appeared to soar, passing above the broken clouds. After rising above the clouds they appeared to be moving directly away from the earth. The largest had an apparent area of about six suns. It was egg-shaped, the larger end forward. The second was about twice the size of the sun, and the third, about the size of the sun. Their near approach to the surface and the subsequent flight away from the surface appeared to be most remarkable. That they did come below the clouds and soar instead of continuing their southeasterly course is also certain. The lights were in sight for over two minutes and were carefully observed by three people whose accounts agree as to details.

—FROM THE MARCH, 1904, ISSUE OF "WEATHER REVIEW,"
A REPORT FROM THE SHIP USS SUPPLY, AT SEA

We start with the most frequently reported and "least-strange" events: *Nocturnal Lights*, lights in the night sky. These represent the major class of reports that I, as an astronomer, had been asked, since 1948, to explain whenever possible as astronomical objects and events.

It should be clearly understood that *initial* light-in-the-night-sky reports have a very low survival rate. An experienced investigator readily recognizes most of these for what they are: bright meteors, aircraft landing lights, balloons, planets, violently twinkling stars, searchlights, advertising lights on planes, refueling missions, etc. When one realizes the unfamiliarity of the general public with lights in the night sky of this variety, it is obvious why so many such UFO reports arise. Of course, such trivial cases do not satisfy the definition of UFO used in this book. Equally, when a UFO *is* defined, as was the case in the Condon Report, as "any sighting that is puzzling to the observer" rather than as we have here—a report that *remains* unexplained by technically trained people capable of explaining it in common terms—one can recognize the reason for the basically unsatisfactory nature of the investigation concerned.

In the Nocturnal Lights category, in particular, we should admit for consideration only those cases reported by two or more stable observers, in which the reported behavior of the light and its configuration and overall trajectory are such as to preclude by a large margin explanation as a simple misperception of natural objects.

After such a critical assessment is made, to dismiss such highly selected cases as being without merit or potential significance for physical or behavioral science is, at best, cavalier and irresponsible.

THE REPORTERS

Since the observer who reports the UFO event is pivotal to any study of UFOs, let us first consider the forty-one reporters concerned in the cases I have selected to delineate the primary characteristics of this category. I suggest that they can, and should—because of their evident qualifications as competent witnesses—be taken seriously by scientists.

The average number of reporters in the selected Nocturnal Lights case was 3.5; the median number was three. Among the thirty-seven adult observers we note a wide range of occupations[1] and technical competence—ranging from a butcher and three housewives to a Royal Canadian Air Force telecommunications officer, a US Naval security officer, and an MIT laboratory head—but most of the observers at the time of their sightings were holding positions of responsibility: pilots (4), air control operators (8), police and security officers (5), etc.—positions in which we would be distressed to find persons who are mentally unstable or prone to silly judgment or hoaxes. In all cases, the reporter observed in concert with at least one other responsible adult.

As we have already noted, often the reporters' immediate reactions, in their own words, can be very enlightening.

In the first category, Nocturnal Lights, we can well start with the reactions of the associate laboratory director at the Massachusetts Institute of Technology. (See Appendix 1, NL-I .) When his eleven-year-old son ran into the house calling, "There's a flying saucer outside," he and the rest of his family went out to look. In our interview the father said:

> Going out of the house, I got my small glasses [4x30] to observe the object. I really didn't believe I was going to see anything. In the meantime, my fifteen-year-old boy went back into the house and got the Bausch and Lomb 6x30 binoculars. We both observed the object.
>
> My very first impression was . . . is it an extremely bright star? But that thought was dispelled almost immediately. The second thought—searching for a logical explanation—was that it might be a landing light of an aircraft. [This theory was soon dispelled by the strange trajectory of the light, as plotted by the observer against the bare patches of a tree. It was midwinter.] . . . the next morning, I asked my oldest boy to describe his observations to me, and these checked with mine entirely.
>
> I don't honestly see how I could call it an aircraft. Besides, I had both the plane and the helicopter for comparison. [These had passed by during the twenty-minute observation period.] Oh, my wife said maybe it was a satellite. I said how could a satellite possibly go through the motions that this did.

Eight airport tower operators figure in the set of Nocturnal Lights cases recorded in this chapter. A comment such as the following—backed up by four other witnesses—deserves attention:

"I've been working in the tower for twenty-seven years, and I've never seen anything like this before. It was the violent maneuver . . . and the apparent cooperation between the two bright objects that made the sighting significant." (See Appendix 1, LN-2.)

Of another UFO sighting reported by an airport tower operator, the witness said: "I've been an air traffic control supervisor for the last four years. I am familiar with burn-ins and satellite crossings. I have tried to figure out what I saw and explain it to myself." (See Appendix 1, NL-3.)

If the observer could not explain it, neither could the Air Force. An official communique commented: "In view of the experience and reliability of the observers [air control operators], it is concluded that a phenomenon of some sort was observed, but the logical cause cannot be determined."

For a change of pace (and occupation) in the matter of immediate reactions to the experiencing of a UFO event here is the comment of a young but mature antiques dealer: "as I kept saying, 'What can it be?' he [her husband] just kept repeating, 'Oh my God'" (See Appendix 1, NL-4.)

On a lighter note, we have the following report:

> One night back in 1961 I was engaged in the noble American tradition of "parking" with a girl. What caught my attention, and at that time it took an awful lot to distract me, was the way the thing [a bright Nocturnal Light] moved. . . . The object was noiseless and, not to sound corny, glowed. It was much brighter than any star in the sky . . . So as it moved slowly northward, I figured it to be a weather balloon reflecting the sun's light. However, balloons don't stand still, change direction, and have reverse gears, so to speak. Well, I finally pointed it out to the girl to assure myself that it wasn't an illusion. She saw it with no trouble and got quite scared. We watched together as the thing went through its antics . . . Finally, after some five minutes of fooling around, it took off for greener pastures. From far to the south it moved out of sight to the north in about five seconds. I timed it, I know it. I don't expect you to believe it, but it happened. [See Appendix 1, NL-5.]

One could quite literally fill a book with such spontaneous reactions of mature observers at the time of their experiences, but it would serve little purpose save that of amplification. Thus we will look at only one more reaction to a UFO in this category:

> Allow me first to give you a bit of information on myself so that you can see that I am a reasonably qualified observer. I am forty-four years old, have been a member of the Canadian Air Force for over twenty-five years, first as a member of air crew during the Second World War. For the last twenty years I have been employed in the telecommunications field. I have spent over half of that time on flying bases and have seen most of the aircraft of both military and

> civilian types . . . I should add that I have never been a believer in UFOs before, but this one is so unexplainable by our present standards that it has me wondering . . . none of the flying experts from the base have an explanation for it either. [See Appendix 1 NL-6.]

THE REPORTS

Turning to what these mature persons reported, let us start with a report transmitted to me by Dr. David Layzer (but not originated by him), of the Harvard College Observatory. In his covering letter Dr. Layzer stated: "Here is an absolutely reliable eyewitness account [eight observers] of mysterious moving lights seen . . . by a neighbor of mine [a member of the faculty of the Harvard Medical School] and several members of his family." (See Appendix 1, NL-7.)

In his letter the doctor stated:

> The object caught my attention because . . . the light looked wrong for an airplane. We often see, from our house, planes with their landing lights on an approach to Logan Airport; usually, however, when I see landing lights I can also see red and green wing lights. In this case it was not possible to see any lights. There was no sound whatever as the object seemed to get closer. . . . It was an exceptionally clear, cold, and still night. . . . When the object appeared to be at its nearest point, I would guess one-half to one minute after it first appeared, a second light appeared on essentially the same course as the first, and my curiosity was further heightened when a third light appeared about a half minute after the second. I immediately went indoors for my field glasses.
>
> Upon returning, I saw that all three lights were still visible; the first two had stopped about 15 to 25 degrees above the horizon and were near to each other and motionless. The third light was still moving. With field glasses no red, green, or other normal running lights could be seen. At this point the lights came, I would guess, about one-half minute apart, a series of them, to a total of six or seven. I am neither a trained observer, nor at the beginning was I particularly trying to keep careful account of what was happening.

> Several of the early lights became completely motionless, while others were moving over the horizon; finally two, or perhaps three, of them from the motionless position appeared to drop smaller lights, which twinkled or flashed as they dropped vertically, and as this happened, the motionless lights appeared to dim and extinguish.

This reporter disclaims being a trained observer. Would that the average UFO report were as coherent and detailed as this one from an "untrained" observer! He continues:

> . . . one of the most striking things about the lights was their color. It was orange light and therefore unlike any I have ever seen on an airplane. Not a vivid or harsh orange but simply too orange to be a normal landing light. . . . During the time when the lights were visible, several planes passed within audible range, but their sound faded and the lights continued with no sound that we could detect. . . . Their speed would certainly be impossible to judge as we could not tell how far away they were or even guess at it. As far as angular speed they moved [between] the same rates as satellites [and] a jet with its landing lights on during an approach to the airport, we very commonly see.
>
> . . . The lights were as bright as Venus as seen at its brightest; that is, very striking lights, but they certainly cast no light on the ground. Subsequent conversations with friends always seemed to result in two questions. . . . First, the lights that moved up from the south toward the northeast were completely steady. They did not twinkle, they did not flicker, they were as steady as the light of Venus or an aircraft landing light. I could see no shape or form or anything else attached to these lights. The lights that appeared to be dropped or detached from the object did twinkle as they fell.
>
> . . . in relistening to this account, it seems to me that the order of events is not clearly stated. . . . I had been across the street at a neighbor's house and was walking back to ours when I saw the first light. My wife was still at their house. About three or four of the lights had appeared, and I had already gotten my field glasses from the house when my sister, her children, and my parents arrived back from church. Even though three of the objects were in the sky, I was still feeling extremely skeptical that this was anything out of the ordinary, although I was extremely curious. In calling these lights to the

> attention of the party that had just arrived, I felt more than a little foolish, and all treated it as something of a joke [a common reaction]. We all passed the field glasses around and agreed that we could see nothing particularly different with or without the field glasses. . . . I went into the house to call Dr. David Layzer, who is a neighbor. Receiving no answer, I came outside . . . [They] were still watching the lights, and the count had become confused. We think that a total of six or seven appeared. The entire episode took perhaps twenty minutes before the last light disappeared from sight. The lights that dropped the little lights were, as far as I could tell, stationary. They were definitely not moving perpendicular to our line of sight. It was easy to keep them centered with the glasses propped against a tree, and their illumination was so steady that I am quite certain they were not vanishing into the distance along our line of sight.

I corresponded with the observer a year later. In answer to my questions, he wrote:

> I would say, yes, the event still seems as strange now as it did then. . . . My own (admittedly unsatisfactory) explanation was that the lights were connected with some type of ordnance work that the public was not supposed to know about. I confess to being open, but essentially skeptical, about extraterrestrial objects and visitors. . . . I have enclosed a carbon of a letter from Donald Menzel [Harvard astronomer], to whom David [Layzer] also sent a copy of my account. I confess I didn't answer his note because, aside from the fact that he seemed to treat the whole matter facetiously, he obviously had not read the report with any care. . . . I would say that his final explanation [bright stars in the main, with an airplane landing light or two, possibly plus a satellite] is out of the question in that, by trying to apply a combination of these objects that he suggests, for six or seven objects that behaved essentially identically, he simply taxes the imagination too much.

The reaction of Blue Book was similar—and negative. When I proposed that an inquiry be made through military intelligence channels as to whether there were indeed any classified exercises being conducted on that cold winter night, my suggestion was met with a complete lack of enthusiasm. Since a consultant has no authority, the matter rested there.

The above sighting is certainly one of the "less strange" variety; possibly it has a normal explanation. I have given it in some detail here first, because it so excellently illustrates the attitudes of some scientists and of Blue Book and second, because it also gives the lie to the contention that only status-inconsistent people report UFOs.

Now, if no sightings involved any greater exhibition of speed, maneuvering, or other indications of an esoteric means of propulsion, we might very well not have a problem. Yet there are cases with great Strangeness Ratings, and, therefore, the sighting quoted above is retained as a UFO because it meets the definition of UFO: the airborne lights and their trajectories remained unidentified by persons deemed capable of identifying them if they were indeed identifiable as a normal occurrence.

Among the dozen or so cases under discussion in the present category, we have the following example. This was recounted by one of the two observers involved, an MIT graduate engineering student:

> At this time Ursa Major [the Big Dipper] was almost at the zenith. I suddenly noticed that two of the stars were moving . . . in a circle about a common center while maintaining position at opposite ends of a diameter, much like two paint dots at opposite ends of a spinning phonograph record. They were rotating about 30 rpm counterclockwise at a very constant velocity. . . . The rotating stars were separated by a distance approximately equal to . . . about one and a half moon diameters. The objects were fainter than Arcturus, a little fainter than Alpha, Beta, Gamma in Ursa Major. . . . abruptly stopped their motion, and this left them in a roughly north-south orientation. . . . They remained dead motionless, they started moving away from each other, the one moving south suddenly halted. . . . The "star" that had begun moving northward continued to do so. At this time its velocity was constant and slower than most meteors but faster than ordinary aircraft. [See Appendix 1, NL-8.]

The case was reported to the National Center for Atmospheric Research at the University of Colorado (not to the Condon Committee) on the advice of two MIT professors, one of whom was the reporter's graduate adviser. It was also reported to the Harvard College Observatory. In neither case was there any follow-up.

The sighting occurred in May, 1970, quite some time after the Condon Committee had concluded that there was no point to further study of UFOs. One can as easily use the following paraphrased excerpts from the taped interview with two policemen who reported a Nocturnal Light case at the time the Condon Committee was just beginning its work. (See Appendix 1, NL-9.) The case was not studied by the committee.

The policemen observed a large, bright, round, white object 50 degrees above the horizon and apparently located between two neighboring towns (as attested by radio reports from these and other locations, which made a rough triangulation possible). The object hung motionless for about fifteen minutes, blacking out when the officers shined their spotlights up toward it. They said it was the size of a silver dollar held at arm's length.*

Shortly afterward a smaller object—a light—streaked in toward it from the northwest, moved close to the bright object, and stopped. Then another light streaked in from the southeast and also stopped close by the large light. Then the large light executed a "square" trajectory, sending occasional blue shafts of light toward the ground. After some thirty minutes of such maneuvering the small lights shot off at high speed in the direction from which they had come, taking about five seconds to disappear. No sound was heard.

Unfortunately the interrogator did not obtain as full an account as he might have, and I did not discuss the case with him until much after the event. But here, as in other instances, we come directly to the question that any serious investigator must ask himself over and over again: how does such a report originate? Either the police officers had for more than an hour been bereft of their reason and were reporting sheer fantasy, and the police-radio operators in the adjoining towns had succumbed to hysteria and were unable to separate facts from fancy as they talked with their colleagues, or these police officers *did indeed* observe something extraordinary.

---

* This is undoubtedly an exaggeration—a very common one in UFO reports. People do not realize how large an angle a silver dollar would subtend on the sky when held at arm's length. Virtually no one realizes, for instance, that an aspirin tablet held at arm's length will cover the moon.

The policemen were not as articulate or learned as the doctor who reported the strange lights seen outside Boston or the MIT graduate student and his wife who reported the whirling starlike lights, but the taped interview indicated that they were certainly equally puzzled.

If it should be concluded that the first hypothesis is the more probable—that the observers were temporarily bereft of their reason—then in view of the many strange reports from police officers throughout the nation (and in other countries), perhaps we should call for a thorough revision of our method for selecting police officials. One would indeed be in a sorry plight if such misguided and nonobjective officers were to testify against one in court. How could their testimony be trusted?

Could it be that pilots are similarly affected by loss of judgment? In the Nocturnal Light category of UFOs, as an example of one of many cases in the files, we have, from a Blue Book "Unidentified" report, the following statement (See Appendix 1, NL-10):

> A reddish-white, blurred, large, luminous glare appeared ahead and 500 feet below aircraft on a collision course. It maintained its altitude but made a right turn when the aircraft commander took evasive action [the report of an Air Force major, lieutenant, and two crew members]. Investigation to date offers no indication of possible causes.[2]

In an official report from an Atlanta-based Eastern Airlines captain, dated February 28, 1968, and made available by an Eastern Airlines flight director, we find this interesting passage:

> I picked up the mike and asked, "Who's this at our 11:30 position?" The center replied that the airplane he was talking to was fifteen miles away. I said, "Well, this guy isn't fifteen miles away."
>
> With this I prepared to take evasive action. The center advised that they still had no target showing, and I said, "Aw, come on! He's going right by us at our nine o'clock position."

It should be remembered that while these are merely two examples from a very great many pilot reports, pilots are wary of making such reports unless they are under military instructions to do so.[3]

Turning now to airport tower operators, whose judgment we citizens trust many times a day for their ability to recognize a plane coming in for a landing and to distinguish between a landing light, Venus, or some "unknown craft," three of the eight tower operators included in the roster of reporters of selected cases of Noctumal Lights concurred in the statement of one of them (there were just three in the tower at the time):

> The two objects [in a deep blue twilight sky, moon present but stars not yet visible] were just bright points of white light and could have been taken for satellites except for the sudden maneuvers, change of direction, and speed of disappearance. . . . One was headed north at 45 degrees above the horizon, the other south at about 30 degrees. The southbound light executed a sudden 180-degree tum, rose, joined the other object, hovered in what appeared to be a formation, and then flew off to the northeast. [See Appendix 1, NL-2.]

The speaker, a tower operator of twenty-seven years experience, was sufficiently impressed to call me long distance to report it. He had four witnesses, two of whom told me in a personal interview during my stay in North Dakota, the scene of the sighting, that they had contacted Great Falls radar and that the presence of an erratic target had been telephonically confirmed. This statement was officially denied the next day, thus adding to the host of reported Air Force and Federal Aviation Administration denials made a day or so after a reported radar confirmation.

Another good example in the Nocturnal Lights category is the "MIT case" because of the unimpeachable qualifications of the principal observer, a man thoroughly acquainted with scientific procedures. The following direct quotations from my taped interview with him refer to the description of the object sighted rather than to his reactions. (See Appendix 1, NL-1.)

> It was much brighter than Venus. It appeared as an intense white—maybe with a slight yellowish tint—source, probably not a pinpoint source.
>
> I would describe it just as a very small source in a very hot furnace, as a central source, white hot type of flame, and then with this peripheral color dancing around on the outside of it, the red and

green—the red bordered on the pink. The other thing we observed as we looked at the object through some small trees [denuded]. It was quite evident that there was a wandering motion of the object with respect to the background of the trees. . . .

Q: How long did it stay in the hovering, wandering position?

A: Somewhere between five and ten minutes.

Q: When we talked about it before, you said something about it being an eerie kind of thing you had not experienced before. In fact, I think you said it was a sort of "radioactive" kind of thing. Can you go into that a little more?

A: I don't know why I said that except that the source was extremely intense, and it was of a color you would not expect to see generated by artificial means such as a lamp—or any known type of lamp.

Q: How would it have compared with a short circuit of electrical wires such as occurs in an ice storm?

A: There would be some similarity there except for the fluctuations of color. The central light was much more steady than you would experience in a thing like that.

Q: Do you suppose it could have been an experimental craft of some sort trying out strobe lights? Did it bear any resemblance to a strobe light?

A: No, it did not.

Q: Now let's go back. Was there any identifying sound?

A: None. None whatever.

Q: What about its later motion?

A: After observing the object for some five or ten minutes in its apparent hovering position and its wandering, it started to increase its altitude and travel toward the east; I would estimate its altitude went up to about 30 degrees, and it arrived at an azimuth of approximately 160 degrees [southeast], at which time it appeared to stop and hover again. This motion, although it did not seem to be in proximity to it, seemed to be coincidental with the passing of an airliner.

Q: I think this sort of reviews our previous discussion. I can't think of any salient facts we left out. Let's try to get the angular rate. We haven't gotten that down. When it was moving its fastest—apparent motion—how would you . . .?

A: It was going somewhat, I would say, in excess of a degree per second. Something of that order of magnitude.

A former chief scientist of the Pentagon, my personal friend and friend and colleague of the MIT reporter, had asked me to look into this case in the first place, calling me from across the country at the time. Despite this instigation from a highly placed professional man, I was unable to get Blue Book to investigate further.

I include yet another Nocturnal Light case because of the circumstances surrounding its reception. After this book was virtually completed, I had addressed a letter to the editor of *Physics Today*,[4] soliciting UFO reports from scientifically and technically trained persons. The following Nocturnal Light case was one of the first responses I received. It is noteworthy in another respect: the report is eleven years old; the reporter, who today is a professional astronomer, did not wish to report it earlier because he was unwilling to expose himself to ridicule.*

This Nocturnal Light sighting took place in Canada. (See Appendix 1, NL-11.) The reporter and his brother had been alerted by a relative, a newspaperman, who, in turn, had been called by the provincial police, who had been attempting to follow the light with their cars but had not succeeded in catching up to it as it moved from place to place. The call had come about 2:00 A.M., after the chase had been on for nearly an hour.

I quote directly from the report, but names and places are not given, (as I promised in my *Physics Today* solicitation). (See Appendix 1, NL-11.)

---

* Another respondent, also a professional astronomer, wrote: "... being a scientist, I had never reported. . . ." This person had preferred to regard his sighting as being of an unusual physical phenomenon rather than to admit the possibility, perhaps even to himself, that it was "a genuinely new empirical observation."

We followed country roads until we came within 100 yards of the object.* It was hovering around a large tree, which stood alone in the center of a cultivated field. The tree was about 100 yards distant and about 120 feet high. The object, which subtended an angle of about 1/4 degrees (giving it a physical diameter of less than 3 feet), appeared circular in shape and was thus probably a spheroid. It was highly luminous against the dark sky background and changed color through the whole visible spectral range with a period of –2 seconds (rather an irregular period). Because it was rather bright, I may have slightly overestimated the angular size, and 1/4 degrees should perhaps be considered an upper limit. A lower limit would certainly be 1/8 degrees.

The object appeared to be examining the tree rather closely. It circled the upper branches, ranging from 50 to 100 feet off the ground, passing *in front* of the tree, then clearly visible *through* the branches on passing behind the tree again. It continued this apparent "observation" of the tree for several minutes while we watched. Then, anxious for a picture, we climbed the perimeter fence and started slowly toward the tree facing due west. We had not gone more than 10 feet before it "noticed" us and, noiselessly accelerating at a very high rate, headed almost directly south, disappearing over the horizon on a *slightly* rising trajectory in about 2-1/2 seconds. (I consider my length and time estimates to be quite reliable as I was actively engaged in track and field at the time and thus quite competent at this type of estimation. Even under such exceptional circumstances, these figures are most probably within 20 per cent.)

Several observations about the object:

1. It was certainly too small to contain human life;
2. It had no apparent physical surface features apart from the circular shape it presented—possibly because the "surface" was highly luminous;
3. It moved deliberately and purposefully in its "inspection" of the tree, pausing slightly at apparent "points of interest" and giving the distinct impression of "intelligent" behavior;

* Because of the distance between the reporter and the object this case falls within the upper limits of a Close Encounter and might be considered as such.

4. Its motion was completely silent, even the final rapid acceleration;
5. It was definitely not any natural physical phenomenon I have ever encountered or read about (I'm sure you are familiar with what I refer to—"marsh gas" and the like);
6. It was definitely not a distant astronomical object. It was clearly visible alternately through the branches of the tree and obscuring the branches of the tree, fixing its distance quite exactly;
7. It was definitely seen by competent witnesses (including several police officers) besides myself;
8. On acceleration from the tree it almost certainly should have exceeded the speed of sound. There was no acoustical disturbance whatever. (My uncle attempted to take a picture of it as it accelerated, but the result was not good enough to publish due to our excessive distance from the object and its rapid motion, which combined to produce a very faint blurred image.)

The salient points to consider are these: the object appeared to be governed by some intelligence, and it did not behave as would a physical phenomenon as we understand it.

The small estimated linear size of the last Nocturnal Light is unusual.† The general impression given by reporters of these cases is that the light is considerably larger than three feet. Since, however, these are nocturnal sightings and only rarely is it possible to judge distances with any confidence, linear sizes remain unknown.

It would be difficult to estimate how many good cases of Nocturnal Lights a diligent investigator might be able to collect. Thousands upon thousands of raw, unfiltered initial reports of Nocturnal Lights very probably exist; how many of them would survive the filtering process and be admitted into the arena of truly puzzling cases remains a matter of conjecture until serious investigation is undertaken. However, the prototype of the Nocturnal Light is clear.

---

† A Nocturnal Light case in Fargo, North Dakota (February 26, 1967), which I personally investigated and was totally unable to explain, involved a light of estimated size of a few feet. (See Appendix 1, NL-12.)

The typical Noctumal Light is a bright light, generally not a point source, of indeterminate linear size and of varying color but most usually yellowish orange, although no color of the spectrum has been consistently absent, which follows a path not ascribable to a balloon, aircraft, or other natural object and which often *gives the appearance* of intelligent action. The light gives no direct evidence of being attached to a solid body but presumably may be.

As far as trajectories and kinematic behavior are concerned, despite exceptions that defy normal physical explanations, even when generous allowance is made for exaggeration and error of judgment, the reported motions of the Nocturnal Lights do not seem generally to violate physical laws.

The thirteen cases used in this chapter are representative of many hundreds of others, by no means agreeing in details but generally faithful to the prototype gleaned from these selected cases. Even were we limited to this handful of cases, it would be most difficult to say that each of them must have been the result of some unusual but natural event, for in not one instance has that "unusual but natural event" been tracked down and established. Some will ascribe this failure to the fact that in none of these cases was a truly in-depth investigation undertaken. (Would that in even a few cases Blue Book had adopted the investigative attitude and procedures of the FBI!) We are left in doubt; we click our tongues and say, "Strange—but there must be some natural explanation."

If so, what is it?

CHAPTER SIX

UFOS Seen in the Daytime—Daylight Discs

A large airplane body with no wings is the nearest I can explain; or perhaps like the outer edge or circumference of a disc rolling towards me, the edge 51 or so feet or more.

—DESCRIPTION OF SIGHTING OF FEBRUARY 4, 1966, IN HOUSTON, TEXAS, FROM BLUE BOOK FILES

In this observational category—reports of UFOs seen in the daytime—we deal primarily with discoidal or oval shapes. There are fewer reports of daytime than of nighttime sightings; even when we limit ourselves strictly to well-investigated baffling cases—true UFOs—we still come up with more nighttime than daytime cases.

Perhaps the UFO phenomenon is intrinsically nocturnal. If it is, there are still many hundreds of good daytime sightings on record. In my own files, filtered daytime entries eligible for the select group do not run far behind the hightly selected nighttime reports, but this may be because I place very high demands on nighttime sightings for inclusion in the file.

THE REPORTERS

Since in approaching the daytime category we must once again start with the observers and their qualifications, I have, as before, chosen a dozen or so representative cases, each of which had at least two reporters.

The total number of witnesses in these daytime cases is sixty; the average per case is 4.8, and the median, four. Many "spectacular" single-witness cases might have been included, but I have felt it wise throughout to omit single reporter cases even when the credibility rating of the person in question is high.[1]

Once again the words of some of the reporters involved in the thirteen cases throw an interesting light on the whole phenomenon. The quotations are all taken from the Daylight Disc cases listed in Appendix 1. It is the reaction and not the substance of the case in which we are interested here. These are on-the-spot reactions to the sighting of a Daylight Disc.

> . . . my friend, who was driving the vehicle, said, "Do you see what I see?" . . . This odd-looking object looked like a stunted dill pickle. We agreed we didn't know what it was. While we stopped there, a half-ton truck came along with two men in it—they were taking a load of hogs into Calgary. The one man asked us if we were having trouble. We said no but showed them the object and asked what they thought of it. One of the men said, "Oh that must be one of those flying saucer things. . . ." However, I sat around and thought about it all day, and that afternoon I decided to phone the control tower at Calgary Airport to see if they knew anything about it. They said they didn't. [See Appendix 1, DD-3.]

> I wish now that I had taken more pictures as it moved groundward in a controlled approach, but I was anxious to see it with the naked eye rather than through the viewfinder. [See Appendix 1, DD-3.]

> I have been an airline pilot for nearly five years and have reasonable vision, and naturally I am used to observing things in the sky. This was not a fleeting glimpse. While I was watching, explanations occurred to me and were discarded on the spot. [From a report by a BOAC pilot of a daylight sighting on July 13, 1971, Kent, England. See Appendix 1, DD-4.]

> During World War II I was a pilot in the US Air Force. In all that time I never once, day or night, observed anything unusual in the skies. Now, at age forty-three, I have observed a phenomenon which is beyond my comprehension and which taxes my sense of reasoning and credulity. [See Appendix 1, DD-2.]

The daytime reporters evince the same reaction of surprise and bewilderment shown by reporters of nighttime sightings. One might well think that in the clear visibility of brilliant daylight several observers simultaneously would not long remain puzzled by a sky sighting, especially when the duration of the event is relatively long. But they do, and generally they try, in vain, to fit some natural explanation to the experience. As well trained as some of the witnesses concerned in these sample cases are, it is surprising to note how often they felt inadequate to put into words a cogent description of their experience.

THE OBJECTS

We can start with those who are perhaps the least technically trained of our present roster of reporters, two farmers who found themselves, at 7:25 A.M., near Three Hills, Alberta, Canada. (See Appendix 1, DD-1.) The best description of the object they reported that they could muster was that it "looked like a stunted dill pickle'." The recipients of this unique description, the drivers of the hog-carrying truck, described the object thus:

> The color was greenish blue. It seemed to have a sort of fluorescent glow to it, but it wasn't really a fluorescent color as we would know it. I would say it was more like the writing on these signs along the highway that say, "Calgary (so many) miles," something like a scotch light with a green background. Actually no [definite] lights on it whatever.

Untutored as these men might be, they certainly would be capable of a more articulate description of lights with which they were familiar, such as lights on cars or on barns. As it is, they tried hard to describe the color of the glow of this "stunted dill pickle" that traveled along with them, following the rise and dips of the hilly land.

Their puzzlement is not unique. Repeatedly I have bad witnesses tell me, "I just can't describe the color. I've never seen anything quite like it before. I never saw just that shade of red [or blue or green] before." Frequently the object is described as having a general fluorescent glow with no specific lights, as in another two-observer case, in which one witness stated, "The outline was definite, but there were no port lights on it to make you think it was a kind of airship or anything like that. No exhaust or jet flames, actually no lights whatever [except for the general glow]."

Descriptions that lack precision of terminology are by no means confined to untrained observers. The same groping for words to convey to the listener a faithful picture of what the observers are sure they saw occurs also in cases involving welltrained observers. Thus the best the two airport tower operators on duty and a third airman on duty at the alert pad at the end of the runway could do was "two oblong-shaped devices having the appearance of a table platter." Yet that morning the weather was clear and cool, and the visibility was excellent.* (See Appendix 1, DD-3.)

Interrogation of many reporters has convinced me that the vagueness of their descriptions (which might appear as a deliberate attempt to confuse the issue and thus to prevent exposure of a misperception, of which the reporters secretly think they might be guilty but to which they are committed) is actually the result of the high Strangeness Rating of the sighting. The reporters simply have a vocabulary inadequate for the situation. I have found that the witnesses seem to be doing the best they can. Farm workers can give accurate descriptions of something with which they are familiar—a tractor, for example, or other farm machinery. A similar ambiguity of expression plagues reporters with considerable technical training: police officers (who are supposed to be able to give accurate descriptions of accidents and crimes), airport tower operators, scientists, engineers. Perhaps the ingenuous description of the conveyors of the load of hogs is the most practical and pragmatic after all: "Oh, that must be one of those flying saucer things!"

It has been my experience also that reporters are usually almost as hard-pressed to describe the *sounds* made by the sighted object. Almost always they say, "That wasn't exactly it, but that's about as close to it as I can come." Daylight disc sightings are, almost without exception, noiseless, and this is reported to be the case from all over the world. Thus in the Calgary case (as

---

*The message sent to Dayton from the local air base stated: "In view of the fact that three reliable personnel reported the sighting . . . it is concluded that a genuine sighting of a phenomenon of some sort did occur but that sufficient information is not available to determine the cause." To the best of my knowledge, no attempt to obtain more information was made after this message was received by Blue Book.

in countless others.): "There was not a bit of sound, but we could hear the sound of the airplane taking off from the airport at Calgary [much farther away]." (See Appendix 1, DD- 1.)

Turning now to the trajectories and kinematics of the Daylight Discs, it is reported that the UFOs' actions generally appear controlled except that frequently a wobbling or tumbling, or "falling leaf," motion is described. The discs appear to have a universal ability to take off smoothly, often with fantastic accelerations and usually without producing a sonic boom.

Newton's Second Law of Motion rules out extremely rapid accelerations for bodies of appreciable mass. It is not my aim, however—here or at any point in this monograph—to pass physical judgment; that requires more data than presently exist in recorded form. I am merely playing the role of the assessor of experiences reported by people good and true, and reports of high strangeness from reporters of high credibility rating do exist. That much is incontrovertible.

Part of the high Strangeness Rating arises from the reported trajectories. Here is an example from a taped interrogation. (See Appendix 1, DD-1.)

> Q: There is some hilly country in there. Did the thing float right along evenly over the hills, or did it follow it?
>
> A: That was one thing we noticed. As high up as it was [500 or 600 feet], it didn't have to do the things it was doing. Whenever there seemed to be a slight rise in the land it made a slight rise. When there was a dip in the land, it seemed to dip. This was another thing I couldn't figure out.

The question was asked deliberately because of my previous experience in such interrogations. The disc "hugs" the contour of the ground over which it glides, often stopping over small bodies of water.

The paradigm of this class is contributed to by other sightings in our selected multiple-witness cases:

> Very briefly, what I saw was a small silvery white disc of unknown diameter, unknown altitude, but definite physical existence; it first appeared stationary, under visual observation, for about ten minutes. Then it moved across the sky, visually passing under the clouds and finally disappearing into the white clouds. No sound could be detected.

> The white dot stood still too long and moved too silently to have been an aircraft; it appeared to travel in a direction distinctly inconsistent with the direction of the clouds so as to preclude . . . that it was a balloon. [See Appendix 1, DD-4.]

The descriptions of daylight sightings are remarkably similar: oval or discoid white or silvery objects, apparently solid. Sometimes a disc is reported to have a dark band along its circumference. "It was like a silvery hamburger sandwich," said a professional sculptor whose report is not included for consideration in this chapter because it was reported by only one person. This disc, or "silvery hamburger sandwich," reportedly executed a large square in the sky and then streaked away "like a frightened rabbit." In another single-witness case, the reporter, a mechanic, used the term "sandwich," with the central rim of the craft described as the edge of meat protruding beyond the slices of bread.

Photographs of reported daylight discs are readily available, and while the circumstances under which they were taken have not been sufficiently investigated and many are patent fakes, it is difficult to dismiss others. Some photographs that I have examined may be authentic daylight disc photographs, for I have not been able to find any evidence of trickery in these cases. Since some celebrated hoaxes have been accompanied by photographs—it would seem hoaxers subscribe to the idea that a picture is worth a thousand words—I am extremely wary of any photograph submitted to me. In my opinion, a purported photograph of a UFO (particularly a Daylight Disc) should not be taken seriously unless the following conditions are satisfied: (1) there were reputable witnesses to the taking of the picture who sighted the object visually at the time; (2) the original negative(s) is available for study because no adequate analysis can be made from prints alone; (3) the camera is available for study; and (4) the owner of the photograph is willing to testify under oath that the photograph is, to the best of his knowledge, genuine, that is, that the photograph is what it purports to be—that of a UFO. The last condition need not apply if the photograph in question is accompanied by several independently taken photographs, preferably from significantly different locations.

Clearly these conditions are stringent, but they must be—usually a photograph is no more reliable than the photographer.* Even when all the conditions are met, all one can say positively is that while the probability that the photograph is genuine is very high, certainty cannot be established. Still, if, for example, twenty-five such instances can each be accorded a very high probability, the compound probability that photographic proof of UFOs exists would be all but indistinguishable from certainty.

I do not know of twenty-five such cases, but there are several that meet nearly all the necessary conditions. One is the classic Great Falls, Montana, case of August 15, 1950 (see Appendix 1, DD-5), in which movies of two point-like lights, in a bright daylight sky, were taken, incorporating a sufficient number of reference objects (for instance, a water tower) to enable a meaningful study of the series of frames to be made. The attempt to ascribe the recorded parameters of the motion of the objects to aircraft, balloons, etc., was entirely unsuccessful. Dr. Baker, writing in the *Journal of the Astronautical Sciences*, concluded:

> Because of the conflict between every hypothesized natural phenomenon and one or more details of the hard data in the photographic evidence analyzed (in addition to the uncertainty of the soft data) no clear cut conclusions . . . can be made. . . . A number of other films have been viewed by the author, which purport to be UFOs, and they all seem to exhibit the common quality of poor image definition. . . . Most of them have been taken with amateur equipment like the Montana film. Like the Montana film, some of these films definitely cannot be explained on the basis of natural phenomena (others can be explained if one searches one's imagination).[2]

I have examined many purported photographs. Most of them are of little scientific value (the object is too distant, no frame of reference, image blurred, etc.) even if genuine, and many lack the quality of conviction. Perhaps the best

---

* The same may be said of radar photographs. Here it is a question not of fakery but of interpretation (assuming the proper fuctioning of equipment) by the operator. So again we are reduced to fallibility of the human element. When all is said and done, the UFO remains a human experience and must be evaluated as such.

that I have personally investigated at some length, which essentially met the criteria listed above, is a photograph of Daylight Discs. It is not mentioned as proof of the existence of Daylight Discs but as the best Daylight Disc photograph I have personally investigated. Even so, not all the circumstances surrounding the taking of the photograph are as clear as they might be.

In this case I was able to obtain the two original negatives and, with the permission of the owner, to subject them to laboratory tests in which the standard lacquer was removed, negative copies were made, and a study was then made by microscope and by flying spot scanner of the grain structure of the original negatives.[3]

In addition to the study of the negatives, the camera was examined[4] and tested, the three witnesses—one of whom, the photographer, was the owner of the camera—were interviewed, and affidavits were obtained from two of them.

The results of the tests leave no doubt that real images exist on the color photographs and that the images satisfy the stated time sequence and the light conditions under which the pictures were reportedly taken (there are no telltale inconsistencies in shadows, cloud movements, etc.). Of course, the real image could be that of a large platter tossed high into the air and photographed. (I say large because a close object would not exhibit the "softening" effect the atmosphere produces when an object, particularly a shiny one, is viewed from some distance.)

To satisfy myself that the locale of the sighting was indeed in the bush and not at all easily accessible for the staging of a hoax I arranged to fly over the specific area in a small plane. It was truly rough, hilly brush country—the foothills of the Canadian Rockies—but not impassable. To mount a hoax at that point would have required monumental motivation, including, I should think, a very good prospect of financial gain.

My repeated conversations and correspondence with the principal observer, Warren Smith of Calgary, have failed to produce any substantiation for such motivation. Smith's affidavits, made under the stringent provisions of the Canada Evidence Act, further support my feelings.

Although the purpose of these chapters is to construct prototypes of the major observational categories of UFOs rather than to present detailed accounts of individual sightings, a synopsis of the Warren Smith sighting will be to the point.

Warren Smith and two of his companions, who prospect as a hobby, were returning from a weekend prospecting mission when, at about 5:30 P.M. on a fairly clear July day, the youngest of the three, a teenaged boy, drew his companions' attention to what at first everyone thought was a plane in trouble. No noise was heard, so they thought that the engines had been cut off. As soon as it was apparent that the object had no wings and was gliding smoothly downward, the men abandoned the airplane hypothesis.

Even before this, however, Warren Smith, who remembered that he had a loaded color camera in his pack, called excitedly for it and started photographing. He thought that the object was an aircraft heading for a crash, and it crossed his mind that the photograph could be sold to the newspapers on their return. (This was the only time in which the idea of monetary gain entered their thinking, as far as I could gather.) One picture was reportedly obtained as the object came down toward the trees in the foreground, behind which the object soon disappeared. Then, the men reported, the object reappeared from behind the trees and ascended toward the clouds. The observers also reported that the object dropped some material, but this report was never fully substantiated.

The entire incident took some twenty-five seconds. The only tangible evidence we have are the two color photographs—taken, unfortunately, with a fixed focus camera—both of which contained real images and gave no evidence whatever of having been tampered with.

The remote possibility exists that quite independently of Smith and his companions and without their knowledge, someone in the bush had at that moment launched a platter, which Smith was fortunate enough to have been on hand to photograph. Yet we have both the word of Smith that the disc was first seen to descend and *then* to ascend and disappear into the clouds and the established sequence in the negatives, which shows that the stipulated descending photograph was taken first. One could argue, even then, that the invisible plattertosser had tossed twice and that Smith photographed the descent of the first one and the ascent of the second, some fifteen seconds or so later, but we have the word of the reporters (who in this case must have been independent of the tossers) that this was definitely not the case. In any event, close examination of the cloud structure shows that the two photographs were taken in close succession; even a brief interlude would have resulted in minor but detectable changes in the cloud edges. None is perceivable.

The Smith photographs portray quite well the archetype of the Daylight Disc, and most descriptions of reporters in the other multiple-witness cases included here support the Smith photographs in this respect. Going as far back as 1952, we have this description of a Daylight Disc from two personnel at the Carco Air Service hangar adjoining the southeast comer of the Kirtland Air Force Base in Albuquerque, New Mexico. (See Appendix 1, DD-1.)

> There appeared high in the sky directly over Kirtland Air Force Base an object which first appeared to be a weather balloon, but after closer examination it was determined by the observers to be of a design unfamiliar to them. It was then noted that a similar object of the same design was nearby. The two objects moved slowly to the south . . . making no sound which could be heard by the observers. The objects were of a round, disc-like design and silver in color. Both objects seemed to pick up instant speed and climbed almost vertically. One object continued on a south-southeast course while the other object veered to an almost due east course. The entire observance took place within thirty seconds. The winds were southwesterly at fifteen miles per hour.

It would be easy to explain away this incident by saying that the observers mistook some very close windblown objects, or perhaps by something else. Or would it? The winds were from the wrong direction; the objects disappeared in different directions, climbing *vertically*. It seems most unlikely that all this could have been accomplished and the objects propelled rapidly upward by a fifteen-mile-per-hour south-southwesterly wind.

As in so many other Blue Book cases, no follow-up was, to my knowledge, undertaken. The credibility of the observers (other than the fact that they were airport personnel) or their motivation in making the report or the manner and attitude in which they made it was never established.

Earlier that year, on January 16, at Artesia, New Mexico, a similar event contributed to the paradigm of this class. (See Appendix 1, DD-8.) The report in Blue Book files goes as follows:

> On January 16, 1952, two members of a balloon project from the General Mills Aeronautical Research Laboratory and four other civilians observed two unidentified aerial objects in the vicinity of the balloon they were observing. The balloon was at an altitude of 112,000 feet and was 110 feet in diameter at the time of the observation.
>
> The objects were observed twice, once from Artesia, New Mexico, and once from the Artesia Airport. In the first instance, one round object appeared to remain motionless in the vicinity, but apparently higher than the balloon. [Nothing is said about what the other object did.] The balloon appeared to be 1-1/2 inches in diameter and the object, 2-1/2 inches in diameter (thus the ratio of 3 to 5), and the color was a dull white. This observation was made by two General Mills observers.

Nothing is said about the assumption that the two objects observed from the balloon's launching site and later from the airport were the same pair. Details of this sort mattered little to the Blue Book investigators.

The Blue Book report continues:

> A short time later the same two observers and four civilian pilots were observing the same balloon from the Artesia Airport. Two objects apparently at extremely high altitude were noted coming toward the balloon from the northwest. They circled the balloon, or apparently so, and flew off to the northeast. The time of observation was about forty seconds. The two objects were the same color and size as the first object. [Here it would seem that the first sighting had only one object.] They were flying side by side. When the objects appeared to circle the balloon, they disappeared [momentarily, it is to be presumed, since they later flew off to the northeast], and the observers assumed they were disc-shaped and had turned on edge to bank.

There was no follow-up by Project Blue Book for the following reasons:

> Unfortunately this report was not made until April 5 and did not reach ATIC until April 6. Due to this time lapse no further

> investigation is contemplated. The observers are known to be very reliable and experienced.
>
> Conclusions: "None."

The time lapse was certainly no excuse for the lack of further investigation. Determination of the qualifications of the reporters could certainly have been carried out, even at a much later date. "The observers are known to be very reliable and experienced," is a meaningless statement without further substantiation.

The following year personnel from General Mills Laboratory figured in another UFO report, class Daylight Disc. (See Appendix 1, DD-9.)

> Three research engineers observed a white smoke or vapor trail at 40,000 to 50,000 feet, while tracking a 79-foot balloon at 73,000 feet through a theodolite, Object moved in horizontal flight for approximately thirty seconds at a rate of 10 degrees per 9 seconds (estimated 900 miles per hour) then began vertical dive lasting 10 to 15 seconds. During dive object was visible several times appearing to glow. As object leveled off, smoke trails ceased. Observation was made from roof of General Mills Laboratory,

Further comments on the Blue Book case card were as follows:

> One of the observers is a meteorological engineer and is considered to be completely reliable. The two other sources are also considered reliable. The nearest AC&W [radar] facility was inoperative at the time of the sighting. Two F-86 aircraft were in the area southwest of Minneapolis at the time of observation, but this does not correlate with the UFO. Conclusion: UNIDENTIFIED.

Not included on the card but submitted by observers in the original report was the statement that the object passed below the sun, which was at an elevation of about 25 degrees. There was no sound.

The observers were a former B-17 pilot, now a meteorological engineer, a private pilot with two years of postgraduate work in supersonic aerodynamics, and a development engineer who made observations by naked eye, the others making their observations through a theodolite. The observers jointly stated, "The possibility that the appearance of a dive was produced by the object merely receding into the distance seems unlikely since the speed normal to the line of sight was undiminished in the dive." They also pointed out that there was no sonic boom and that "the vertical dive was a highly dangerous if not suicidal maneuver."

The best attested case of "UFOs appearing to be interested in balloon launches"—to be anthropomorphic for a moment—was reported by my friend Charles Moore, Jr., an aerologist and balloonist, in 1949. (See Appendix 1, DD-10.) Moore described the event to me personally.

He was in charge of a navy unit involving four enlisted personnel; they had set up facilities to observe and record local weather data preparatory to the Special Devices Center Skyhook operation. The instrumentation on hand consisted of a stopwatch and ML-47 (David White) theodolite, a tracking instrument consisting of a 25-power telescope so mounted as to provide elevation and azimuth bearings.

At 10:20 A.M. the group released a small 350-gram weather balloon for observation of upper wind velocities and directions. Moore told me that he followed the balloon with a theodolite for several minutes, after which he relinquished the tracking instrument to a navy man with the admonition "not to lose it or he'd be in trouble." Moore then picked up the weather balloon with his naked eye, and shortly thereafter, looking back at the man at the theodolite, he noticed that the instrument was pointing elsewhere.

Using a few choice navy expletives, Moore was about to snatch the instrument from the man and direct it at the weather balloon when the man said, "But I've got it in here." Moore looked and saw a whitish ellipsoidal object in the field of the theodolite. The object was moving east at a rate of 5 degrees of azimuth change per second. It appeared about 2-1/2 times as long as it was wide. It was readily visible to the unaided eye and was seen by all the members of the group. In the theodolite it was seen to subtend an angle of several minutes of arc.

As it became smaller in apparent size, the object moved to an azimuth reading of 20 to 25 degrees, at which point the azimuth held constant. Coincidentally, the elevation angle suddenly increased, and the object was lost in the telescope. It disappeared in a sharp climb—thus resembling other Daylight Disc cases—after having been visible to Moore and his group for over a minute.

The sky was cloudless; there was no haze. The object left no vapor trail or exhaust. No noise of any kind was heard in connection with the sighting, and there were no cars, airplanes, or other noise generators nearby that might have blotted out sound coming from the object. As the day progressed, many airplanes flew over and near the balloon launching site, and Moore's group was able to identify them by appearance and engine noise. They saw nothing again that day that bore any resemblance to the white elliptical unidentified object. To a man of Moore's training, this was a "real" event. And as later events proved, it was not an isolated case, though, as usual, to the best of my knowledge, it was not taken seriously by Blue Book. No follow-up was made.

Three other cases used here to delineate the Daylight Disc prototype occurred in 1967, a year relatively high in UFO incidents of all types in the United States. The three are listed as Unidentified by Project Blue Book, and occurred in Crosby, North Dakota (see Appendix 1, DD-11), in Blytheville, Arkansas (see Appendix 1, DD-3), and in New Winchester, Ohio (see Appendix 1, DD-I2).

In the North Dakota case there were seven witnesses, five in one family and two observers located 20 miles away. However, the Air Force investigator did not bother to interrogate the two completely independent witnesses, thus losing a chance for getting a geographical fix on the object, for determining its speed and trajectory and for getting completely independent testimony on the nature of the object in question. Of the five witnesses in one location, only one was interrogated, and then only by telephone. What a different situation this might have been had a proper investigation been conducted!

From what information we do have on the North Dakota case, we know that an oval, luminous object reportedly appeared from behind a barn and windbreak, then climbed noiselessly upward and disappeared. Since it was a commercial pilot (and his family) who saw this "apparition," I am unwilling to discount this sighting as a simple misperception.

The incident reported from New Winchester, Ohio, with five witnesses, was also poorly investigated, although listed as Unidentified by Blue Book. The original report was prompted by an article I had written for a friend, then editor of a house organ published by an insurance company based in Columbus, Ohio.[5]

This Ohio case adds its own piece of information to the Daylight Disc prototype:

> This object was oval in shape and was going in a straight line from southeast to northwest in a very much tumbling fashion. . . . The UFO, or whatever it was, crossed over the road we were going on. There was bright sunshine, and it reflected on the object, which was made of metal and was not the color of aluminum like airplanes, but I would say the color of either brass or copper. What powered the object we do not know, but we heard no sound.

Obviously there are not many hard data here, and the incident could easily be dismissed if it didn't fit the pattern of so many other similar reports. There were no trained witnesses in this case, but the letter of transmittal has a frank and open style:

> We were driving east and saw a car with three youths in it, ages about eighteen to twenty years, stopped and they were looking at something in the sky. . . . I saw something, too, so I pulled on down the road a little ways and stopped, got out of the car, and looked in the sky. . . . The three boys . . . came on down the road and parked beside my car and we were all watching it, and the traffic came from the opposite direction, and we both had to move. . . . None of the five of us had any explanation, but we all saw it very plainly.

After one has had the experience of interrogating many observers and of reading many letters and reports (and has also had ample chance to meet and interrogate *bona fide* members of the lunatic fringe), one would be obtuse indeed if one did not develop a feeling about narratives that have the ring of genuineness about them as contrasted with those that are the products of maladjusted minds. The sincerity and the puzzlement of many witnesses are beyond question.

In Blytheville, Arkansas, two observers on duty in the control tower at Blytheville Air Force Base and a third observer on duty at the south end of the runway (all three observers were considered by Blue Book to be "completely reliable") saw "two oblong-shaped devices" having the appearance of a table platter. The objects, dark against the sky but with an exhaust of approximately seven feet, were sighted suddenly from the control tower. Their estimated altitude was 1,200 to 1,500 feet. They traveled on a straight line from east to west but disappeared after fifteen to thirty seconds during a turn to the southwest.

The report stated that the visual spotting was "confirmed by Blytheville Air Force Base, RAPCON, as being some two nautical miles distant." This has not been established as a Radar-Visual case because of the lack of specific radar data.

An Air Force official stated: "This is the first phenomenon of this kind reported in the vicinity of Blytheville Air Force Base for which there has not been a ready explanation." He continued, "In view of the fact that three reliable personnel reported the sighting . . . it must be concluded that a genuine sighting of a phenomenon of some sort did occur but that sufficient information is not available to determine the case."

Blue Book was content to list this case as Unidentified without further investigation. This is understandable to some extent in view of the inadequate staff of Blue Book.

My call for good UFO reports published in *Physics Today* produced a good multiple-witness Daylight Disc case. (See Appendix 1, DD-14.) Reported by a professional astronomer, it occurred in 1965, though for obvious reasons the observer hesitated to report it.

The sighting was made just after sunset, but the clear sky was still starless. The report stated that the object had lights as well as a disc shape, indicating, perhaps, that here we have a transition case between the Nocturnal Light and the Daylight Disc. The trajectories and kinematics of the two categories are strikingly similar, perhaps suggesting that Nocturnal Lights are Daylight Discs seen at night and that, therefore, the distinction between the two categories is purely observational.

The observer, accompanied by his wife, her friend, and two children, was traveling eastward at 30 miles per hour. They noticed "a silvery, disc-shaped object heading slowly south. The bottom of the object had a ring of bluish-white lights, which made the object appear to rotate." The object subtended an angle of two to three degrees and was topped by a white light.

> After moving slightly to the south of us, the object rapidly accelerated in an east-northeast direction. We moved onto a high-speed highway still heading east, but now at 70 to 80 miles per hour. The object quickly became a white starlike object [Nocturnal Light] far to the east of us. It appeared to move five to ten degrees up and down for about five minutes. Then the object rapidly moved to the south, disappearing over the Atlantic Ocean.

PROTOTYPE

On the basis of these reports, we can now summarize the salient features of the Daylight Disc.

Those that I have investigated to any degree are characterized by similarity in shape, in color, and particularly in their manner of motion, which can be extremely slow—even hovering close to the ground or executing a slow pattern of motion—or extremely rapid, so that the disc can disappear in a matter of seconds.

Despite the presence of daylight in the dozen cases used in our prototype, all we really glean from them is that the object (often objects in pairs) is variously described as oval, discshaped, "a stunted dill pickle," and ellipsoid. It generally is shiny or glowing (but almost never described as having distinct point source lights), yellowish, white, or metallic. It exhibits in most cases what we would anthropomorphically describe as "purposeful" directed motion, with the ability to accelerate extremely rapidly. No loud sounds or roars seem to be associated with the Daylight Discs; sometimes there is a faint swishing sound.

The sad fact is that even after years of reports of Daylight Discs from various parts of the world, and despite some seemingly genuine photographs, the data we have to deal with are most unsatisfactory from the standpoint of a scientist. Part of the reason for this is clear: official apathy and the "ridicule gauntlet."

The majority of the reports of Daylight Discs with which I have spent any time came from people of at least some training, of established common sense, who are reasonably articulate. Yet the desired details, so necessary for any meaningful study, elude us. Why? In very large measure simply because no one in authority (and in the United States this means the Air Force) conducted any investigation worthy of the name.

What investigations were carried out (and I overheard many phone conversations during my regular visits to Blue Book) and what questions were asked were almost always aimed at establishing a misperception, and the questions were so directed. Rarely were the questions set in the framework of "Here may be something quite new; let's find out all we can about it. What were the details of its trajectory (never mind if they did seem to violate physical law)? Describe as best you can exactly what happened first and what happened next. How much time did it take to do that part of its motion, how many times a second did it wobble, how many seconds did it take to cover an arc of twenty-five degrees?" Are these questions that the average observer cannot answer? Nonsense. Given patient interrogation (rather than the desire to fill out a form quickly) of a normal person, one can by reenacting the crime—preferably at the scene of the sighting—obtain such a time-motion sequence even if the investigator must translate the observers' words, "It took as long as it takes me to count to ten to go from above that tree to the edge of the barn" into, "The object had an angular rate of two degrees per second."

Colors can be checked by the use of a good color wheel (I never came across a Blue Book investigator who used one); and brightness, "As bright as that yard light over there" can be translated into lumens and finally into a rough estimate of ergs per square centimeter even if only the upper limits to the distance can be ascertained, as is the case in which the luminous source passes in front of an object (tree, house, hill) the distance of which is known.

But investigations conducted in that manner were notoriously absent in Blue Book procedures. Investigations were predicated on the assumption that all UFO sightings were either misperceptions or the products of unstable minds. Such official failings are tragic in the extreme, though, as we have seen, not uncommon. Examination of another set of cases, those which involved both radar and visual evidence, makes this point clear.

CHAPTER SEVEN

Radar-Visual UFO Reports

At about 1040Z ECM operator No. 2 reported he then had two signals at relative bearings of 040 and 070 deg. Aircraft cdr. and co-pilot saw these two objects at the same time with same red color. Aircraft cdr. received permission to ignore flight plan and pursue object. He notified ADC site Utah ADC site Utah immediately confirmed presence of objects on their scopes.

—FROM OFFICIAL REPORT OF WING INTELLIGENCE OFFICER

On the surface it would appear that instances involving both radar and visual mutual confirmation of a UFO should offer superior hard data. Yet such is unfortunately not the case. A lack of follow-up and the application of the Blue Book Theorem—it can't be, therefore it isn't—prevented, in my opinion, such hard core data from being properly reduced and presented.

Radar sightings of UFOs might appear to constitute hard data, but the many vagaries of radar wave propagation are such that it is almost always possible to ascribe a radar UFO sighting to such vagaries if one tries hard enough. Nevertheless there are in Blue Book files examples of radar sightings that carry the classification Unidentified (even though one Project Blue Book chief officer testified before Congress, but not under oath, that there was no radar UFO case in the Blue Book files that had not been satisfactorily explained[1]). Unidentified radar cases in Blue Book are, for example, the sightings of September 13, 1951, at Goose Bay, Labrador (see Appendix 1, RV-10); August 13, 1956, at Lakenheath, England (see Appendix 1, RV-4), from which report the quotation heading this chapter is taken; June 3, 1957, at Shreveport, Louisiana (see Appendix 1, RV-6); and December 6, 1952, in the Gulf of Mexico (see Appendix 1, RV-11).

It is often stated that UFOs are not picked up on radar. It is quite true that, as far as has been officially disclosed, the highly mission-oriented radar defense coverage of the country does not appear to yield a crop of UFO observations. UCTs—*Uncorrelated Targets*—are observed on the North American Radar Defense (NORAD) radar screens, but since these do not satisfy the conditions of a ballistics trajectory, they are automatically rejected without further examination. It would have been an easy matter to introduce a subroutine into the NORAD computer system that would isolate the UCTs without interfering with the basic mission of NORAD; but despite my suggestion to that effect, Blue Book never adopted the idea. Consequently it is not possible to state that reports in which radar is involved are intrinsically rare. It may be that while they are not officially reported, they are by no means rare.

In any case, radar sightings are reported. When visual observation accompanies a radar UFO sighting and when, of course, the visual and radar observations can be established as definitely having reference to the same object or event, there is great promise of scientific paydirt if proper investigations are made.

As before, I have chosen a dozen or so representative cases (listed in Appendix 1) to illustrate this category and to construct a prototype displaying the overall pattern of the sightings, using, as before, direct quotations from the observers. In addition to the human experience, we have added an instrumental experience, which gives strong support to the former.

THE REPORTERS

We start, also as before, with the types of reporters involved in this category of cases.[2] In no case are there fewer than two observers for any one sighting reported. The average number of witnesses is 5; the median number, 4.5. Ten of these cases are from Blue Book files, only two of which are officially listed as Unidentified. One case that Blue Book has listed as Anomalous Propagation the Condon Report lists as "Unidentified." Blue

Book has evaluated the remaining seven cases as probable aircraft (4); possible aircraft (1); aircraft, mirage, and radar inversion (1); probable balloon and probable aircraft (1). None of the Blue Book identifications has been substantiated by positive evidence, largely, perhaps, because in none of the cases was there adequate follow-up.

The reactions of various observers to their experience are interesting. A pilot and his student had been informed by the tower that radar showed a UFO on their tail for the past five minutes. The pilot acknowledged the report, stating that the object was not a conventional aircraft. The pilot said: "We were more petrified than anything else as to what it was. Maybe it was going to shoot us down for all we knew." (See Appendix 1, RV-12.)

THE REPORTS

The following excerpts from a transcript of a conversation between a Lear jet pilot, the Albuquerque control tower (see Appendix 1, RV-1), and a National Airlines pilot are revealing with respect to both reactions and attitudes.

Prior to the excerpts given, a conversation had been in progress between the Albuquerque control tower and the pilot of a Lear jet near Winslow, Arizona. The jet had been describing a red light, initially at their ten o'clock position, that flashed on and off and that quadrupled itself in a vertical direction. The Albuquerque radar "painted" just one object whenever the light was on, none when it was off. The light repeated the quadrupling process a number of times, seeming to retract into itself the lights below the original light; then as the tower warned the jet that the object was getting closer, it seemed to play a cat-and-mouse game with the jet, involving some rapid accelerations.

After some twenty-five minutes and with terrific acceleration, according to the jet pilot, whom I interviewed at length but who insists on anonymity, the object ascended at a 30-degree angle and was gone in fewer than ten seconds. The Albuquerque radar, according to the jet pilot, painted the object until the time of its final acceleration and disappearance. A brief portion of the radio conversation involving the Lear pilot (L), the Albuquerque tower (A), and a National Airlines pilot (N) is revealing of both reactions and attitudes.

A to N: Do you see anything at your eleven o'clock position?

N to A: We don't see anything.

A to N: Are you sure nothing at your eleven o'clock position?

A to N: Did you hear conversation with Lear jet?

N to A: Yes, we have the object now—we've been watching it.

A to N: What does object appear to be doing?

N to A: Exactly what Lear jet said.

A to N: Do you want to report a UFO?

N to A: No.

A to L: Do you want to report a UFO?

L to A: No. We don't want to report.

Another representative Radar-Visual case, illustrative not only of Radar-Visual cases in general but also of the operation of the Blue Book Theorem, involved two commercial airlines pilots and an Air Traffic Control Center operator. (See Appendix 1, RV-2.) Blue Book dismissed the case as "landing lights" on the word of a reluctant American Airlines pilot, who clearly did not wish to get involved. I received a letter from the air traffic controller, who answered my inquiry for further information thus:

> I have pondered on whether to make a reply to your letter. . . . However, the more I thought about the explanation the Air Force gave for the incident, the more disturbed I have become. . . . I have been an air traffic controller for thirteen years, three actual years of control in the US Air Force and ten with the FAA. What happened on May 4, 1966, is as follows: I was assigned the Charleston, West Virginia, high altitude radar sector on the midnight shift. . . . At approximately 04:30 a Braniff Airlines Flight 42 called me on a VHF frequency of 134.75 and asked if I had any traffic for his flight. I had been momentarily distracted by a land-line contact, and when I finished (ten to fifteen seconds), I looked at the radarscope and observed a target to the left of Braniff 42, who was heading eastbound on jet airway 6, about five miles off to his eleven o'clock position.

I advised Braniff 42 that I had no known traffic in his vicinity but was painting a raw target off to his ten o'clock position; however, it was not painting a transponder and was probably at the low altitude sector (24,000 feet and below). Braniff 42 advised that the object could not be at a low altitude because it was above him and descending through his altitude, which was 33,000 feet. . . . I was completely at a loss for explanation for I advised him [that] at the time there were only two aircraft under my control—his flight and an American Airlines flight about 20 miles behind him. I asked Braniff 42 if he could give me a description of the object, thinking it might be an Air Force research aircraft or possibly a U-2 type vehicle. Braniff 42 advised that whatever it was, it was not an aircraft, that the object was giving off brilliant flaming light consisting of alternating white, green, and red colors and was at this time turning away from him. At the same time the American flight behind the Braniff, who had been monitoring the same frequency, asked the Braniff if he had his landing lights on. Braniff advised the American negative. Even if Braniff 42 had had his landing lights on, American wouldn't have seen more than a dull glow, for they were twenty miles apart and going in the same direction! Which means to me that the American saw the same brilliant object. When I asked the American if he could give me any further details, he politely clammed up. Most pilots know that if there is an official UFO sighting, they must (or are supposed to) file a complete report when getting on the ground. This report, I understand, is quite lengthy.

I contacted Braniff 42 and said I saw this target come at him from about eight to ten miles at his ten o'clock position and at a distance of about three miles, make a left turn, and proceed northwest bound from the direction it had come from. Braniff 42 confirmed this and added that it was in a descending configuration at about twenty degrees off the horizon.

As I have stated, I think my previous experience speaks for itself, and I know what I saw; and I'm sure the pilot of Braniff 42 was not having hallucinations. The target I observed was doing approximately 1,000 miles an hour and made a complete 180-degree turn in the space of five miles, which no aircraft I have ever followed on radar could

> possibly do, and I have followed B-58s declaring they are going supersonic, all types of civilian aircraft going full out (in the jet stream), and even SR-71 aircraft, which normally operate at speeds in excess of 1,500 miles per hour.
>
> Doctor, that concludes my statement. I am forwarding a diagram showing the geographic location of the jets and the object.

Conflicting evidence was given by the American Airlines captain in a letter to Project Blue Book:

> I did not place any significance to the incident, and to me it only appeared to be an airplane at some distance, say six or eight miles, who turned on his landing lights and kept them on for three or four minutes, then turned them off.
>
> I asked the radar operator if he had a target at my nine or ten o'clock position, and he replied that he did not have, and I said, "Well there's one there all right." I had no idea he was going to turn in a UFO report. I thought nothing further of it. I presume it was the Air Force refueling. I still think it was just an airplane with its landing lights on.

The air traffic controller's testimony, combined with that of the Braniff captain, is *consistent*, whereas the American Airlines pilot's sketchy statement is not. It is inconceivable that an Air Force refueling mission, which involves at least two maneuvering planes, would be in progress six or eight miles ahead of an airliner on a commercial jetway. A refueling mission invariably shows a great many lights. Why would American ask Braniff whether he had his landing lights on, especially when Braniff was miles ahead of him and facing the wrong way? Further, both Braniff and the controller placed the object at Braniff's ten o'clock position and thus ahead of the Braniff, which itself was twenty miles ahead of the American.

Yet American did say he saw something at his ten o'clock position, and if brightness caused American to misjudge the distance and place it much closer to him, hence apparently behind Braniff, this still would not account for the ten o'clock position. Again, if it was some dozen miles behind Braniff, why ask Braniff if he had his landing lights on?

Since Project Blue Book seized on the testimony of the American Airlines pilot and did nothing to follow up this case by obtaining depositions from the air traffic controller, from Braniff, and from American, this case and many similar to it do not constitute scientific data, and little can be proved by them.

All that can really be said of the Radar-Visual cases is that, in a number of instances, responsible persons at radar posts and at visual posts (air traffic controllers, pilots, etc.)—posts requiring responsible attitudes—agreed that highly puzzling events were simultaneously detected visually and by radar. But what were the exact time-motion sequences, the exact trajectories, accelerations, the detailed nature of the radar blips, and to what extent did the several observers agree on details? All these factors remain distressingly unknown and will continue to do so in future Radar-Visual cases (and in other categories) unless the subject of UFOs is accorded scientific respectability, and thorough investigations are allowed to be carried out in a responsible manner.

Insofar as a prototype of the Radar-Visual case is concerned, it can be said that the radar operator observes a blip on his screen that, he avers, is definite, is akin to the type of blip given by a large aircraft, is not the result of malfunction, and does not resemble weather phenomena. A visual sighting is characteristically a light, or possibly a formation of lights strikingly unfamiliar to the observer, with generally only a suggestion, if that, of an object dimly outlined by the brightness of the lights. The speeds involved are invariably high, but combinations of high speed at one time and hovering at another are not uncommon. Reversals of motion and sharp turns, not abrupt 90-degree turns, are characteristic of Radar-Visual cases.

Virtually all Radar-Visual cases are nighttime occurrences, a point that might be considered as damning evidence against the reality of the targets. But we are examining the data and evidence as reportedly experienced by the observers, not as we preconceive it ought to be seen. In the Close Encounter categories daytime sightings do occur with considerable frequency.

An interesting example of a Radar-Visual case that contributes to the prototype and illustrates the cavalier disregard by Project Blue Book of the principles of scientific investigation occurred in New Mexico on November 4, 1957 (see Appendix 1, RV-3), just prior to the celebrated Levelland, Texas, Close Encounter cases (Chapter Eight). The officer who prepared the report, a lieutenant-colonel in the Air Force, said of this case:

> The opinion of the preparing officer is that this object may possibly have been an unidentified aircraft, possibly confused by the runways at Kirtland Air Force Base. The reasons for this opinion are:
>
> 1. The observers are considered competent and reliable sources and in the opinion of this interviewer actually saw an object they could not identify.
> 2. The object was tracked on a radarscope by a competent operator.
> 3. The object does not meet identification criteria for any other phenomena.
>
> That is, the observers were reliable, the radar operator was competent, and the object couldn't be identified: therefore it was an airplane. In the face of such reasoning one might well ask whether it would ever be possible to discover the existence of new empirical phenomena in any area of human experience.

The report of this incident in the Blue Book files is as follows:

> SOURCE'S DESCRIPTION OF SIGHTING: At 050545 Z November [10:45 P.M. local time], both SOURCES were on duty alone in the control tower at Kirtland Air Force Base, New Mexico; this tower is slightly over 100 feet high. One of the controllers looked up to check cloud conditions and noticed a white light traveling east between 150 and 200 miles per hour at an altitude of approximately 1500 feet on Victor 12 [a low altitude airway]. SOURCE then called the radar station and asked for an identification of the object. The radar operator reported that the object was on an approximate 90-degree azimuth from the observer; it disappeared on 180-degree azimuth from the tower observer. The object angled across the east end of runway 26 in a southwesterly direction and began a sharp descent. One SOURCE gave a radio call in an attempt to contact what was believed to be an unknown aircraft that had become confused about a landing pattern. A LOGAIR C-46 had just called in for landing instructions. The object was then observed through binoculars and appeared to have the shape of "an automobile on end". This was estimated to be fifteen to eighteen

> feet high. One white light was observed at the lower side of the object. The object slowed to an estimated speed of fifty miles per hour and disappeared behind a fence at Drumhead, a restricted area which is brilliantly floodlighted. This is approximately one-half mile from the control tower. It reappeared moving eastward, and one SOURCE gave it a green light from the tower, thinking it might be a helicopter in distress. The object at this point was at an altitude of 200 to 300 feet; it then veered in a southeasterly direction, ascended abruptly at an estimated rate of climb of 4,500 feet per minute, and disappeared. SOURCE stated the object climbed "like a jet," faster than any helicopter. (SOURCE estimated this rate of climb.)
>
> Although there were scattered clouds with a high overcast, visibility was good. Surface winds were variable at 10 to 30 knots. SOURCES observed the object for five to six minutes, approximately half of which was through binoculars.

The Air Force officer who prepared the report stated:

> Both SOURCES, interviewed simultaneously, made identical replies to all questions, and gave identical accounts of the sighting. Both appeared to be mature and well-poised individuals, apparently of well above average intelligence, and temperamentally well qualified for the demanding requirements of control tower operators. Although completely cooperative and willing to answer any questions, both SOURCES appeared to be slightly embarrassed that they could not identify or offer an explanation of the object which they are unshakably convinced they saw. In the opinion of the interviewer both SOURCES are considered completely competent and reliable.

Meanwhile, what did the radar operator—physically separated from the visual observers—indicate that he saw on his scope? The following teletype message indicates that the agreement with the visual sighting was excellent except in the manner of disappearance of the object. The visual observers stated that it ascended abruptly in a southeasterly direction; the radar report has the object finally disappearing in the northwest, some ten miles from the radar station. Possibly there is an inconsistency in this, but, equally, the radar may have had the object on its scope considerably longer than the visual observers had it in sight. The radar report states:

> Observer was called by tower operator to identify object near east end of east-west runway. Object was on an approximate 90-degree [east] azimuth from the observer. Object disappeared on 180-azimuth [south] from observer. Object was first sighted on the approximate east boundary at KAFB [Kirtland Air Force Base] on an east-southeast heading, where it reversed in course to a west heading and proceeded to the Kirtland low-frequency range station [was this the same as Drumhead?], where object began to orbit. From the range station object took northwest heading at high rate of speed and disappeared at approximately ten miles from observer.

The radar report adds something not noted by the visual observers:

> About twenty minutes after disappearance [of the unknown object] an AF C-46 4718N took off to the west, making left turn out; at this time observer scanned radar to the south and saw the object [presumably the same unknown] over the outer marker approximately four miles south of north-south runway. Object flew north at high rate of speed toward within a mile south of east-west runway, where he made an abrupt turn to the west and fell into trail formation with the C-46. Object maintained approximately one-half mile separation from the C-46 on a southerly heading for approximately fourteen miles. Then object turned up north to hover over the outer marker for approximately one and one-half minutes and then faded from scope. Total duration of radar sighting: twenty minutes [as opposed to the four to five minute visual sighting].

What, indeed, can one say of a Radar-Visual case like this? The basic agreement of the radar and visual reports and the competence of the three observers, in my opinion, rule out questions of mirages, false returns on radar, etc. *Something* was quite definitely there. If it was an ordinary aircraft, one must ask how it was that the two visual observers, with a total of twenty-three years of control tower experience, could *jointly* not have been able to recognize it when visibility conditions were good. Even if there were no radar confirmation of the slow and fast motions of the object, or indeed just of the presence of an unknown object, this question would still have to be answered. The description of the object's appearance through binocular "like an auto standing on end"—would also demand explanation.

The lack of adequate follow-up—apparent inconsistencies in the radar and visual disappearances should have been checked, and a far more detailed documentation of the entire incident likewise should have been undertaken—plus the application of the Blue Book Theorem led inexorably to misguided aircraft as the only possible solution for Project Blue Book.

While they lend themselves better to investigation than do UFO reports of the first two categories we have examined, Radar-Visual reports offer a special challenge to the investigator. Two classic cases, investigated in as much detail as was possible after the passage of several years by the late Dr. James McDonald, have been treated in the *Flying Saucer Review*[3] and in *Astronautics and Aeronautics*,[4*] respectively. They need not, therefore, be treated in detail here. One occurred on July 17, 1957, at Lakenheath, England. (See Appendix 1, RV-4.)

The Lakenheath case involved two separate ground-radar operators, one military pilot, and one air control tower operator. It was the subject of grossly incomplete investigations both by Blue Book and by the Condon committee, whose conclusions, however, are worth noting: "In summary, this is the most puzzling and unusual case in the Radar-Visual files. The apparently rational, intelligent behavior of the UFO suggests a mechanical device of unknown origin as the most probable explanation of this sighting." But then common sense comes to the rescue: "However, in view of the inevitable fallibility of witnesses, more conventional explanations of this report cannot be entirely ruled out."

The report does not suggest what conventional explanations might cover the situation. In another section of the Condon Report this case is brought up again, with this unsatisfying statement: "In conclusion, although

---

*The UFO Subcommittee of the American Institute of Aeronautics and Astronautics (AIAA) after publishing their Appraisal of the UFO Problem (November, 1970), in which they concluded that the UFO phenomenon was worthy of scientific study, announced that from time to time they would publish in their journal selected UFO cases so that their readers could form their own judgment of the problem. The Lakenheath case, studied by Dr. McDonald, was one of the cases they chose.

conventional or natural explanations certainly cannot be ruled out, the probability of such seems low in this case, and the probability that at least one genuine UFO was involved appears to be fairly high." Nothing further is stated in the Condon Report or conjecture as to what this "genuine UFO" might be.

Probabilities, of course, can never prove a thing. When, however, in the course of UFO investigations one encounters many cases, each having a fairly high probability that a genuinely new empirical observation was involved, the probability that a new phenomenon was not observed becomes very small, and it gets smaller still as the number of cases increases. The chances, then, that something really new is involved are very great, and any gambler given such odds would not hesitate for a moment to place a large bet.

This point bears emphasis. Any one UFO case, if taken by itself without regard to the accumulated worldwide data (assuming that these have already been passed through the "UFO filter"), can almost always be dismissed by assuming that in that particular case a very unusual set of circumstances occurred, of low probability (but strange things and coincidences of extremely low probability do sometimes occur). But when cases of this sort accumulate in noticeable numbers, it no longer is scientifically correct to apply the reasoning one applies to a single isolated case. Thus, the chance that a thoroughly investigated UFO case with excellent witnesses can be ascribed to a misperception is certainly very small, but it is finite. However, to apply the same argument to a sizable collection of similar cases is not logical since the compounded probability of their *all* having been due to misperceptions is comparable to the probability that if in one throw of a coin it stands on edge, it will stand on edge every time it is thrown.*

* An objection can be raised, and correctly so, that the above argument is specious in that a numerical probability value cannot be assigned to the chances that a given report was not the result of misperception. The analogy is valid only to the extent that one feels justified in saying, as the Condon Report did for one case in particular and implied in several others, that the probability was high that at least one genuine UFO had been encountered and thus that the probability that the sighting was due to misperception was numerically quite low.

The second classic case is summarized in the introduction of the Astronautics and Aeronautics article:

> An Air Force RB-47, equipped with electronic countermeasures (ECM) gear and manned by six officers, was followed by an unidentified object for a distance of well over 700 miles and for a time period of 1.5 hours, as it flew from Mississippi, through Louisiana and Texas, and into Oklahoma. The object was, at various times, seen visually by the cockpit crew as an intensely luminous light, followed by ground radar and detected on ECM monitoring gear aboard the RB-47. Of special interest in this case are several instances of simultaneous appearances and disappearances on all three of those physically distinct "channels" and rapidity of maneuvers beyond the prior experience of the air crew.

A Radar-Visual case that the Condon committee did not examine and of which it was probably not even aware—which Blue Book dismissed as having "insufficient data," though no attempt was made to obtain further data, and as "aircraft"—was reported from a navy ship in the Philippines. The sighting occurred on May 5, 1965. (See Appendix 1, RV-5.) I quote from the official report:

> At 060910, in position twenty degrees twenty-two minutes north, 135 degrees fifty minutes east, course 265, speed fifteen, leading signalman reported what he believed to be an aircraft, bearing 000, position angle twenty-one. When viewed through binoculars three objects were sighted in close proximity to each other; one object was first magnitude; the other two, second magnitude. Objects were traveling at extremely high speed, moving toward ship at an undetermined altitude. At 0914, four moving targets were detected on the SPS-6C air search radar at ranges up to twenty-two miles and held up to six minutes. When over the ship, the objects spread to circular formation directly overhead and remained there for approximately three minutes. This maneuver was observed both visually and by radar. The bright object which hovered off the starboard quarter made a large presentation on the radarscope. The objects made several course changes during the sighting, confirmed visually and by radar, and were tracked at speeds in excess of 3,000 (three thousand) knots. Challenges were made by IFF but were not answered. After

> the three-minute hovering maneuver, the objects moved in a southeasterly direction at an extremely high rate of speed. Above evolutions observed by CO [Commanding Officer], all bridge personnel, and numerous hands topside.

The ship in the Philippines added the following to its report, in defense of its crew as careful observers:

> During the period 5–7 May, between the hours 1800 and 2000, several other objects were sighted. These objects all had the characteristics of a satellite, including speed and presentation. These are reported to indicate a marked difference in speed and maneuverability between these assured satellites and the objects described above.

The report is hardly scientific. One would like to know what were these "extremely high speeds" and how it was that with such high speeds the radar could "hold" the objects for as long as six minutes. Did the six minutes include the three-minute hovering period or not? What sort of blips were observed on the radarscope? What course changes were made and with what angular acceleration? And when the objects "spread to circular formation directly overhead," were they then stationary? Did they wobble or move back and forth? Blue Book should have explored such questions.

The witnesses to Radar-Visual cases are among the best technically trained of those who have reported a UFO experience, yet often their words also portray the same sort of dismay and incomprehension that grips the lesser trained. In the Lakenheath case the radar operator requested the pilot of the Venom Interceptor plane to acknowledge that the UFO had begun a "tail chase" of the fighter, as though to confirm his dismaying observations. The pilot so acknowledged and advised that he was "unable to shake the target off his tail," requested assistance, and remarked, "Clearest target I have ever seen on radar."

An account of a Radar-Visual sighting by the captain of a Trans-Texas airliner (see Appendix 1, RV-6) illustrates not only the prototype of these cases but, once again, the everpresent reluctance, especially on the part of technically trained people, to report a UFO. The member of the 4602d Air

Force intelligence squadron who interviewed the pilot in this case and prepared the report to Blue Book wrote:

> SOURCE was reluctant to talk about object as he was somewhat upset because he was being interviewed on the sighting. He felt that he had nothing to do with originating the preliminary report other than asking the AC&W [radar] site if he had company on his flight. After an explanation by the investigator he became cooperative and should be considered reliable.

There follows in the original Blue Book report:

> SOURCE'S description of the sighting: one object was sighted on takeoff from Shreveport, Louisiana, airport at approximately 2030 CST 3 June 1957. Altitude of object was approximately 400 feet when first sighted. SOURCE stated that the control tower called his attention to the object, which appeared as a small light. Landing lights of SOURCE's aircraft were flashed, on and off, and the object responded momentarily with very brilliant light directed at his aircraft. Object then gained altitude from a seemingly hovering position, at a high rate of speed. At this time another object was sighted at about the same altitude and having the same appearance of the first object. SOURCE stated he then contacted the tower to ascertain whether they had both objects in sight. Tower had both objects in sight, using binoculars. Objects then paralleled course of SOURCE's aircraft, moving at about the same speed, which was approximately 110 knots, only at a higher altitude than that of his own aircraft. At Converse, Louisiana, objects were still with them, so SOURCE decided to call GOATEE [radar site] to see if they had object on their weapon [sic]. An affirmative answer was received. SOURCE compared the size and appearance of objects to that of a star; however, he mentioned that at one time he could see the silhouette of objects but would not make a definite statement to that effect.

When interviewed, the co-pilot fully confirmed the pilot's statement but added that the object was at 1,000 feet and a half mile distant when first noted. He said that the light moved on a course of 170 degrees while rising to an altitude of approximately 10,000 feet at considerable speed, after which it appeared to maintain the same relative position to the SOURCE's aircraft for the next hour. He stated also that the radar site reported that they had two objects at 9,700 feet.

The brief statements of the two pilots and the reported reply from the radar site are unsatisfactory and incomplete and therefore frustrating. Project Blue Book lists the case as Unidentified, but as so many times before and after this incident, the unknown nature of the cause was not a spur to inquiry and assiduous follow-up. The UFO had been satisfactorily identified—as *Unidentified.* With the object in view for approximately an hour, if the report is correct, a detailed and conscientious investigation surely could have determined whether the unknown could possibly have been a misperception of natural objects by both pilot and co-pilot, the tower observers, and, presumably, by the radar operator, although it was never firmly established that the radar was indeed sighting the objects that were sighted visually. Of course, if the radar wasn't sighting the visually sighted objects, what was it observing?

The cases so far described serve adequately to establish the prototype of the Radar-Visual category. *Good* Radar-Visual cases, properly investigated, are rare. Those that do exist, however, cannot be easily dismissed. The case already referred to, involving an RB-47 and described in full in *Astronautics and Aeronautics*, July, 1971, is certainly one that must be considered seriously as illustrating an unquestionably strange phenomenon. (See Appendix 1, RV-8.) It is impossible to discuss the case as the result of a misperception or a radar malfunction or as an effect of anomalous propagation. This Radar-Visual encounter occurred on July 17, 1957, while a special electronics plane flew through Mississippi, Louisiana, Texas, and Oklahoma. It was by no means a localized event of short duration; it involved ground and air crews and several radar installations.

Earlier that same year, on February 13, 1957, a challenging Radar-Visual case occurred at Lincoln Air Force Base, in Nebraska. The Blue Book summary reads:

> Objects were visually observed by three control operators and by the Director of Operation, who was in town to supervise a wing mission. Objects were also observed on radar by the NCOIC and

> GCA operation (two separate radar installations). The objects were observed for a period of three to five minutes. . . . The individual objects were about five to six miles behind an air liner and moving twice as fast. . . One of the objects broke in two and another made a 180-degree turn. All observers were interrogated by IFF with no response. Visual estimation of the size of the objects was impossible, but the radar operator stated that the blip on his scope was about the same size as that received from a B-47. The objects appeared to stand still and then speed up and rush away.

Blue Book, applying its standard theorem, evaluated the sighting, "probable balloon" and "probable aircraft."

CHAPTER EIGHT

Close Encounters of the First Kind

Suddenly I realized the light was coming from overhead. I looked up and saw the outlines of an object moving out past the pitch of my roof, approximately 250–500 feet high. The red glow was coming from beneath the object, about center.

—SEE APPENDIX 1, CEI-3

UFO sighting reports that speak of objects or very brilliant lights close to the observers—in general less than 500 feet away—by definition fall into the second large observational division of UFO sightings: the *Close Encounter*. In all likelihood this division does not imply a different order of UFO reports but merely reports of the same stimuli responsible for reports in the first three categories that now, by chance or by design, are seen close up. It is eminently probable that UFOs seen at a distance will sometimes be encountered close at hand, and it is, therefore, purely for convenience in description and study that we make this distinction.

In turn, this large category quite naturally divides itself, operationally and observationally, into three distinct groups: the Close Encounter *per se*, in which the observers report a close at-hand experience without tangible physical effects; the Close Encounter in which measurable physical effects on the land and on animate and inanimate objects are reported; and the Close Encounter in which animated entities (often called "humanoids," "occupants," or sometimes "UFOnauts") have been reported. We have already made the distinction between this latter category and the "contactee" category.

The definition of Close Encounter is best given by the observers themselves, operationally: what are the most frequent distances reported in cases in which the object was close enough to have shown appreciable angular extension and considerable detail, in which stereoscopic vision was presumably employed, and in which fear of possible immediate physical contact was reported? From the reports themselves this appears to be a few hundred feet and often much less—sometimes twenty feet or less. In any event, the reported distance is such that it seems only remotely likely that the actual stimulus could have been far removed, particularly when the object or light passed between the observer and some object (tree, house, hill, etc.) from a known distance away.

It is in Close Encounter cases that we come to grips with the misperception hypothesis of UFO reports. While some brief can possibly be established for this hypothesis in the case of the first major division of UFO reports—those that refer to sightings at a distance—it becomes virtually untenable in the case of the Close Encounter. The UFO reports now to be described, each made by two or more observers who were capable of submitting a coherent, seemingly factual report, raise the question whether the reported perception can possibly be said to fall within the limits of misperception applicable to sane and responsible people.

My own opinion, and I believe the reader will agree, is that accepted logical limits of misperception are in these cases exceeded by so great a margin that one must assume that the observers either truly had the experience as reported or were bereft of their reason and senses. Yet the evidence of the observers' occupations, training, and past performance gives no indication of the latter circumstance in the filtered cases used in this chapter.

Do we then have a phenomenon in which several people suffer temporary insanity at a given instant but at no other time before or after? If so, we have to deal with a new dimension of the UFO phenomenon. But the *data* of the problem—the subject of this book—would remain unaltered.

Simply, the problem of their generation would need to be attacked from another direction.

The same general pattern of treatment of the cases in this category will be followed as in the first large category: UFOs in the sky. First, in each of the subdivisions the number and nature of the observers involved will be stated; second, their firsthand reactions to their experience will be related; and third, the category prototype will be fashioned from elements common to most of the sightings. As before, the individual cases used are listed in Appendix 1.

It must be emphasized that cases I have used here are representative of those that meet the criteria of admission as true UFO reports, that is, reports from responsible people the contents of which remain unexplained in ordinary terms.

CLOSE ENCOUNTERS OF THE FIRST KIND

Close Encounters in which no interaction of the UFO with the environment or the observers is reported can be called Close Encounters of the First Kind. A representative set of these selected from my files are mostly Blue Book cases, and we will examine them for the prototype of this category.

The observers are characterized by the absence of specialized occupations—radar operators, pilots, and air traffic tower operators—that naturally would be present in Radar-Visual observations. We seem to have a more representative cross section of the population as reporters in the Close Encounters of the First Kind category.

As before, I have selected a dozen or so multiple-witness cases from which to build a prototype. The majority of the reporters concerned was interviewed personally by taped phone interviews or by mail. In each instance I satisfied myself that I was dealing with normal and quite sane people and attempted to check one witness against another for consistency. Cogent, coherent reports from single reporters do exist in fair numbers, and in some respects it seems manifestly unfair not to include many of these, for some are of great interest and fit the prototype. Yet for the sake of consistency I have not deviated from the plan adopted in the first three categories.

The cases used here involve forty-two reporters;[1] there were at least two witnesses in each case, the average number being 3.5 and the median number, 3.

Generally, the observers were not independent in the sense that they were located in different places but were independent in terms of background, experience, and, presumably, psychological temperament. They also differed with respect to their previous knowledge of UFO phenomena. In four of the reported events the observers were not physically together and not in communication until later. Vocations of the observers indicate, in many cases, some basic training in critical thought and in the proper discharge of responsibilities: president of a small airline, school principal, and seven police officers, for examples,

In Close Encounter cases it is not easy to separate the reactions of the observers from the description of the event; the two seem to go hand in hand.

A standard question that I have posed to witnesses during the past years is: "If you could substitute some familiar object—a household object or anything that is familiar to you—for the object you saw, what would you choose that had the greatest resemblance, particularly in shape?"

The answer to this question has often been revealing. In one case a witness said, "A beach ball. Just like one of those beautiful beach balls." Another witness, a police lieutenant located several blocks away and presumably viewing the same object from another compass direction, said, "It was like a yo-yo. It was moving off to the northeast. I was sighting it over the top of some trees. It was like a glowing ball—a luminous ball."

This observation was made at 3:00 A.M. The police officer reported that the object hovered and then moved away very rapidly. (See Appendix 1, CEI-1.) A lighted balloon does not satisfy both independent observations, even apart from the fact that it is not very likely that some prankster would be launching a balloon at 3:00 A.M. outside a very small North Dakota town. Nor do the persistently horizontal track, the hovering, the sudden rise at the end satisfy the balloon hypothesis.

The officer continued:

> When I sat there, I had a sort of fear; I wasn't scared for myself but for what it might mean. I sat there, I suppose, for about five minutes. It bounced up and down, like a ball bounces on each word of a song in the movie theater screen, but when it left it was gone—bang—it was out of sight in less than five seconds. It went straight up, right on up In my mind it was guided by somebody or something, like a balloon floating in the air wouldn't have this sort of motion.

Just what sort of a phenomenon are we dealing with?

In a second case (see Appendix 1, CEI-2), involving several witnesses riding together in a car, the principal reporter, a former nurse, answered:

> Well, you know, you have seen these saucers that kids ride down the hill on, you know what I mean? You put two of those together with the rims separating, and I swear it looks just as near that as I can describe anything. . . . I wouldn't say it was reflecting, I would say it was more [self] luminant—you know, like when you look at clock hands that are luminous at night.

Of herself the witness said:

> I have had no military experience, but emergencies often arise in the hospital nursing field, and one must learn to school oneself to maintain composure, which I feel was most helpful to me at the time of our close-range sighting. I worked for twenty-five years as a nurse, and I always try to school myself to be calm and not panic. I think that helped me some.

Continuing her description of the object, she said:

> I know it was something physical. I'll never believe otherwise. . . . I just can't believe it was gas or anything. The outline was very sharp. It was never fuzzy at any time. . . . Then as we watched this possibly for five minutes, it just got a tremendous burst of speed and sped right off. No sound whatever, though. It was something solid, as much as if I were to go out and see an airplane. . . . It was just like looking up under an airplane, just as if an airplane were standing there . . . just perfectly motionless.

A few more direct quotations will help to establish the prototype. It would be so much easier to do this if one could say that all the sightings in a given category had certain things in common—four wheels, windshields, headlights, airplane wings, etc. Yet in fact, the common denominator in sightings such as these, seems to be bewilderment and a universal groping for words of description.

> As I looked out of the window, I realized that the neighborhood was lit up in a red glow. My first thought was that a police car was parked nearby or a fire truck. I called to my wife that something must be wrong in the neighborhood and to come and see. Suddenly I realized the light was coming from overhead. I looked up and saw the outline of an object moving out past the pitch of my roof, approximately 250–500 feet high. The red glow was coming from beneath the object, about center. It appeared as a stream of light coming from inside through a hole. . . . My neighbor's green pickup truck looked brownish. [See Appendix 1, CEI-3.]

Then this witness, to the best of my knowledge quite unacquainted with UFO lore, described an effect reported to me many times:

> An airplane took off from the airport and passed overhead of the object. All the lights went out until the plane was past it. Then with approximately four bright flickers, the object moved from west to southwest and through the overcast. . . . It seemed to me that this object was charting a course or investigating different objects on the ground, as the lights would stop on certain objects such as cars, pickups, hedges, shrubbery, houses, utility lines, and poles.

One can almost sympathize with Project Blue Book officers who took refuge in identifying a case such as this as Unidentified and going on to something else. The above case remains listed as Unidentified in Blue Book's files; no attempt was made even to sweep it away by appending a "possible helicopter" to this case (as was done in others) probably because it would have been too far-fetched even for Blue Book: it *was* 6:oo A.M. on a Sunday morning in midwinter, an unlikely time for a helicopter to be about, even if this interpretation weren't ruled out by the complete absence of noise.

We tum now to another case, involving two Oklahoma farm boys who were stacking hay in the presunrise hours and were taken completely by surprise by the sudden, close appearance of a brightly lighted circular but wingless craft. Excerpts from a rather long taped interview may help the reader to form his own composite picture of Close Encounters of the First Kind and to establish the archetype of this class. (See Appendix 1, CEI-4.)

Q: Did you ever see anything like it before?

A: No. I never did.

Q: What impressed you most about it?

A: The brightness of it.

Q: What do you think it was?

A: I don't know what it was. It scared me at first.

Q: Do you think it could have been a balloon or something like that?

A: No, it was not a balloon or nothing like that. . . . We thought it was helicopters at first from the Quentin Air Force Base, so we called, but they said there were no helicopters up then.

Q: Did they say they had anything on radar?

A: No, they said there were no airplanes or nothing out that night.

Q: Did it have any effect on the animals?

A: Well, the dogs started barking. I didn't notice anything about the cows, but the dogs started barking.

Q: Well, do you think they were barking at it or something else?

A: I don't know, but that was the only thing around to bark at.

Q: You don't know of anyone else who saw it that night?

A: No, I guess there weren't many people up at four o'clock in the morning.

Q: How come you two were up so early?

A: We were hauling hay.

Q: How did you first happen to see it?

[At this point the questions became directed, by phone, to the other witness, in a different part of the country.]

A: He happened to see it first and he came back, and he was scared. I didn't know what was going on.

Q: Did he looked scared?

A: Yeah, he was scared. He was real scared. That's the reason I went out there, to see what he was scared about.

Q: How come you never saw it leave?

A: Well, I thought it was going to crash, and I headed back into the barn too.

Q: Oh, I see. So both of you headed into the barn?

A: Well, yes sir, that's right.

Q: I don't blame you at all. I probably would have been scared too. What color was it?

A: Well, it was just luminous white.

Q: What impressed you most about the whole thing?

A: Well, I guess the fact that it wasn't an airplane. It was some other object.

Q: Have you ever seen anything like this before?

A: Never have.

Q: Would you want to?

A: Now that it happened, I would sort of like to have a picture just to prove that I saw it. A lot of people don't believe me.

Q: How long would you say you were frightened by the thing?

A: Well, it really shook us up for about two weeks. I'd been having trouble getting to sleep. I believe in them now; I didn't before . . . until I saw it.

Q: Have you done much reading about UFOs?

A: I have since then. And I'm going to take some astronomy, here in college.

Q: We're doing our best to try to find out what this is all about.

A: Well, I tell you what, the way these guys acted out here, I thought maybe they had something they weren't telling us about.

Q: You mean the guys from the Air Force? [Air Force investigators were sent by the local Air Force base to investigate.]

A: Yeah.

The craft and its trajectory were described by drawings in the correspondence with the boys. The bright light came down, at a 45-degree angle, to the height of nearby telephone wires, moved horizontally across the farmyard, and was last seen over a small silo. In size it appeared as large as or larger than the full moon. The drawing indicated a circular craft with no obvious protrusions or mechanical features and was described as having "numerous lights around the outside."

The sighting occurred at approximately 4:00 A.M. Sunrise was at 4:44 A.M. local time; hence the sky was by no means fully lighted. One of the teenaged boys stated, "The center of the craft is what has me puzzled, as either it or the whole ship was rotating in a counter-clockwise direction. It was also very shiny in the middle and very, very bright." The entire incident lasted less than three minutes, but under no circumstances could the duration or the trajectory be satisfied by identifying it as a bright meteor. The boys had great difficulty describing in familiar terms what was to them a very real experience—a common difficulty, as we have seen.

Contrary to the general plan of this book, I now offer data obtained by another investigator, Raymond Fowler, an experienced and dedicated observer. The data are taken from a 68-page report prepared by him of a sighting in Beverly, Massachusetts.[2] Fowler, who has undertaken a far more exhaustive investigation of the report than either the Condon committee or Blue Book staff, submitted his full report to Blue Book; typically, they disclaimed any responsibility because the sighting report had not come through official channels. Thus this most interesting case, which the Condon group could not solve, not only was not investigated by Blue Book but was disregarded by it.

The case involves a fairly long-duration sighting of a "luminous platter" that silently hovered over a schoolhouse and that at times approached the reporters so closely that they feared it might crash down upon them.

A few excerpts from the detailed Fowler report must suffice here in contributing to the prototype of this category of sighting. Once again let us go to the taped interviews, for these give us perhaps the greatest insight into the UFO as a human experience.

> This object appeared larger and larger as it came closer. . . . All I could see above my head was the blurry atmosphere and brightly lit up lights flashing (not blinking) slowly around. I was very excited—not scared—very curious. I would not have run at all except for the fact the object got too close, and I thought it might crash on my head,

And from another witness to the same sighting we learn:

> I started to run. Then a friend called, "Look up. It's directly over us"—so I looked up and stood still in surprise. I saw a large round object just at rooftop level. It was just like looking at the bottom of a plate [a familiar pattern]. It was solid . . . I heard no sound at all, but I felt this thing was going to come down on top of me. [It was like] a giant mushroom. I was fascinated, stunned, unable to think, and I automatically found myself running away from it.

One of the police officers who had been summoned to the scene reported:

> At 9:45 P.M. on orders from the station, I went with Officer B to Salem Road, site of Beverly High School, on a report of a UFO. On arrival I observed what seemed to me to be like a large plate hovering over the school. It had three lights—red, green and white—but no noise was heard to indicate it to be a plane. [The duration of the sighting—forty-five minutes—obviously rules out a plane.] This object hovered over the school and appeared almost to stop. The lights were flashing. The object went over the school about two times and then went away.

This was a multiple-witness case, including two police officers, and Blue Book paid no attention simply because it had not been officially reported. The Condon Committee was unable to offer even a tentative natural explanation for the principal sighting, and as for the hypothesis that this was caused by a misperception of Jupiter, Fowler argues convincingly against this interpretation, pointing out among other things that lines of sight established from the interrogation of separate groups showed that the line of sight to Jupiter and to the mean position of the object differed by some fifty degrees. (Of course, it remains possible that some of the supernumerary witnesses may have identified Jupiter as the object after it had receded into the distance, not having noted Jupiter previously under the press of more immediate and local circumstances.)

As far as the paradigm of the Close Encounter sightings of the First Kind is concerned, we may say that the reporters are conscious primarily of a luminous object, sometimes very bright—as intense as a welder's torch—and sometimes merely glowing, like a neon bulb or a luminous dial watch. The shape of the craft seems to be secondary to the luminescence in the perception of the observer, but when a shape is described, it is generally stated to be oval, "football shaped," often with a dome atop it. Rotation of the lights and presumably of the craft is often reported to be in a counterclockwise direction. Hovering is common, as is lack of sound, and very frequently a rapid takeoff without an accompanying sonic boom is reported.

For reports so strange as these the pattern spread is remarkably small. One might expect that hallucinations, for instance, would cover a very wide spectrum. UFO Close Encounters, as reported, do not; there is even a sort of monotony to UFO reports (as UFOs are defined in this book), particularly of the Close Encounter variety. One gains the impression that the differences that exist arise in part from the varying abilities of observers to describe an unfamiliar occurrence.

To add to our concept of the prototype we have the following description of a Close Encounter of the First Kind (see Appendix 1, CEI-6) from a former naval officer. The sighting began when the father, driving his son the officer home from the railway station a little before midnight, saw an object glide in front of them, almost directly over the car. After this had happened three times, he said to his son, "Did you see something glide over the car?"

"Yes, I did," the son answered. "'It looked like a huge prehistoric bird of some kind." When later in an interview I asked my standard question about what familiar object might be substituted for it as far as shape and—in this case—size, the witness said:

> Very hard to say . . . I've never seen anything like it. . . . Well, a navy sub, but not just like that, of course . . . I figured I could hit it with a stone. . . . It was that close. . . . Very sharp . . . just as sharp an outline as if it had been, well, a boiler up there.

The sighting the father and his son were describing lasted for five to eight minutes; the father described it further:

> I dropped my head and looked up through the windshield, and I just looked at it completely—there it was. I said to John, "My God—it's a flying saucer"—it was almost like a science fiction movie on TV. . . . It just hung there, completely silent, like a church steeple lit up at night. Or it looked like those Japanese suicide planes that used to get into the floodlights at night—and this reminded me of that. It swung in an arc of a hundred yards or so—just like it was frustrated.

When two other cars came along the lonely road, they reported that "it" turned off its lights "just like a rheostat on a dining room fixture, and left only a dark shadow" then shot up into the sky, with a trailing blue light after it. The father continued:

> When we got back to the cottage, John said, "Dad, there's something you don't know—when you were at the [car] trunk bending over your camera, this thing moved over the trunk and came down within five yards of you. . . . but I heard no sound."
>
> If John hadn't been with me, I'd have gone to a psychiatrist.

Navy subs, boilers, prehistoric birds, footballs, mushrooms, soup bowls, hamburger sandwiches, and many other analogies—all to describe something that to the observers was essentially indescribable in ordinary terms. These are the sorts of things the investigator hears.

But let us continue, this time with another Unidentified case in the Blue Book files, reported by a school principal and his companions (in another car). (See Appendix 1, CEI-7.)

> I was coming home from a PTA meeting and heading down a small country road, blacktop, and I was thinking of the blackboards the PTA had promised to give me for my school. All of a sudden I noticed a glow coming from over the cliff—and I thought, well, one of the old goony birds [C-47s] is off course, and she's going to land in this cornfield. And this was the first thing that hit my mind. Then this unbelievable object—shaped something like a world War I helmet—came over the top of the cliff. . . . I slowed down at this point. . . . I couldn't understand why an airplane would be on this glide path—and this huge object, over 300 feet, I'd estimate, came over the cliff and stood still almost directly over me for a split second like any object changing direction and then took off towards. the airport. . . . It was terrific bright light. The top of the car seemed to have no effect in holding out light. It was a terrific bright light, unbelievable, I tell myself. When I looked at my hands, it looked like I was looking at X-ray photos.

The principal joined his companions in the other car, which had been following at some distance, and together they watched the object hover over some power lines for about ten minutes.

> Well, then I decided the airport should know about it, so I headed over toward the airport. But I didn't have to tell the people outside. They'd seen it. A couple of lawyers from Kansas City were still standing there with their mouths hanging open. It had flown practically over the airport, but they hadn't told the tower yet. It's a small airport, and there's no glass tower where they'd be watching. They were busy inside because the Ozark flight was due in. . . . By the way, the Ozark pilot . . . if I remember hearing the radio correctly, said, "I see it—it's below me—it's huge," as he was coming in for a landing. . . . When the Air Force came down . . . his [the investigating lieutenant] attitude was not "did you see it," but "how much of it did you see?"

This sighting has remained unidentified to this day. Characteristically, Blue Book did not, to my knowledge, sponsor any sort of comprehensive investigation. Two of the observers, teachers, have preferred to remain incommunicado, and I was able to get a tape interview only with the school principal.

In yet another sighting, far to the north, in Canada (see Appendix 1, CEI-8), the president of a small Canadian air service and his nightwatchman reported:

> It was shaped like two saucers with their open tops touching, one above the other. . . . The entire object was a beautiful silvery white color and appeared to send out rays from its surface, making the object appear like a light on a foggy night.

The executive had gone down to the dock to check the tiedown ropes of his seaplanes. It was the nightwatchman who first called his attention to:

> An object streaking toward us from the west. It was saucershaped and swung and dipped around some low cloud. It kept in the clear and did not enter any of the low scud drifting across the sky. It tilted on its side about 600 feet from us, then straightened out with the flat side parallel with the ground. . . . It stopped dead still in front of us, forty feet above the surface of the lake and about seventy-five yards from us. But distance is hard to judge at night when you are looking at a bright object. No sound came from it, and we could detect no door The thing appeared to me to be only four or five feet across and eight to ten inches thick.

This object appears definitely to have been smaller than similar reported objects, though the difference may be attributed to misjudged distance. Too, there have been other cases in which the smallness of the reported object has been surprising. In any case, that the two men perceived, independently, a strange object and shared what to them was a real experience cannot be seriously doubted. I corresponded at length with the principal reporter, and Brian Cannon, an able investigator from Winnipeg, has made available to me the results of his interviews with both men. On a cloudy night in the northern Canadian lake and woods country what could be "misperceived" to yield the above description?

The Canadian went on to describe his experience:

> It seemed to sparkle as if some electric force or very hot air was flowing from all the surfaces. . . . The machine, after its first stop, slid sideways for a distance of so feet and stopped again. Its speed was not faster than four miles per hour. After about a minute or two

> we could see it accelerate so fast it disappeared like a shrinking star in three seconds from a standing start. Its direction was the same as it came, from the west. Its climbing angle would be about forty degrees. I reported this sighting to the Canadian government. . . . The color was a silvery white. I can't explain the color. I've never seen a color like it. . . . It was bright, but it did not have a glare. It looked more like a fluorescent glow. . . . It was a continuous sparkle like a diamond. It was a bright, beautiful looking thing.

By this time the reader should have some concept of what is reported in a Close Encounter case. But what were the stimuli that gave rise to the puzzlement of the observers?

The obvious sincerity of those who reported UFOs (as defined here), attesting to real events in space and time, stands in contrast to the relatively small number of persons who report a *given* UFO. Why do not more people report specific sightings or, discounting the reluctance to report, why do there seem to be so few people around when a "genuine" UFO appears? It appears to be a phenomenon associated with the absence of large groups of people (there are exceptions, however). It is impossible to establish how many people have seen a UFO but have not reported it or how many sky-observing stations, such as satellite tracking stations, observe UFOs that are never reported.[3]

Obviously there are many unknowns. We must accept the scarcity of UFO observers and reporters as a fact of the total UFO phenomenon, as we do the results of the Michelson-Morley experiment or the fact of the quantum of energy. Like the phenomenon itself, it calls for an explanation and cannot be taken as an argument for the nonexistence of the phenomenon.

One case not only brings to a focus the nature of the Close Encounter phenomenon but also stands on the record as an example of the ludicrous manner in which Project Blue Book sometimes went about investigating a case. A more lucid example of the disregard of evidence unfavorable to a preconceived explanation could hardly be found. Were such blatant disregard of evidence to occur in a court of law, it would be considered an outrageous travesty of legal procedures. The astounding disregard and distortion of reported facts, failure to listen to witnesses, and obdurate and adamant closemindedness can be explained either as incompetence of the most gross variety or as a deliberate attempt to present a semblance of incompetence for ulterior purposes.

The story is one of comedy—of errors, of egregious disregard of testimony, of seeming intrigue, of excitement (involving a car chase at 105 miles per hour), and finally, of tragedy. It deserves to be told in some detail and should someday be published in full. I was involved only peripherally in the affair since I was not called in as consultant until a very late stage, but I watched it develop from the start with great interest. Much credit must go to William Weitzel, instructor in philosophy at the University of Pittsburgh, Bradford Branch, who with care, industry, tact, and persistence brought together the many details of this Close Encounter account. I have Mr. Weitzel's permission to use material from his exhaustive report on the case, containing much personal correspondence with the observers and with government officials.

The case was not examined by the Condon committee, which, indeed, may never have heard of it even though the report was made just a half year before the committee undertook its work. Had it conducted an investigation, I firmly believe another "unknown" would have been added to the substantial number of Condon cases that remain unsolved. In interest, had an unbiased examination of the case been undertaken by the University of Colorado group, it would surely have unearthed some interesting data.

If it were not for the unhappy circumstance that the initial reporter, who took the brunt of ridicule, became a virtual outcast, suffered a disrupted home and marriage, and was made to bear outrageous personal embarrassment, this case history could well be considered high comedy. Three other observers—two of whom were geographically independent of the initial witness and his companion—through the vagaries of press coverage and the failure of the Air Force to interrogate them, escaped the accusation, by implication, of gross incompetence, hallucination, and even insanity—even though they independently described the UFO much in the same manner the spotlighted witness did.

It started out in a very routine fashion. On the night of April 16, 1966, Deputy Sheriff Dale F. Spaur, a full-time member of the Portage County,

Ohio, sheriff's office, after a dinner of steak and eggs, took a two-hour nap, had two cups of coffee, and reported for duty at midnight. (See Appendix 1, CEI-9.) He was immediately dispatched to check a prowler complaint (nothing was found). He received a call to pick up Wilbur Neff, a mechanic who on occasion rode with the regular deputy as a mounted deputy. The two men were then dispatched to answer a call about a car that had sheared a utility pole near Atwater Center, Ohio. They had the driver sent to the hospital and the car towed. Then an Ohio Edison repairman came to fix the pole.

The deputies drove to nearby Deerfield to get some coffee for themselves and to bring back a cup for the repairman. In Deerfield they assisted a man whose car had broken down and arranged to have it towed. They returned to the scene of the pole accident at about 4:45 A.M.

While they were talking with the Ohio Edison man, their police radio reported that a woman in Summit County, directly to the west of Portage County, had reported a brightly lighted object "as big as a house" flying over her neighborhood. The object, the woman reported, was too low to be a plane and too high for a streetlight. Jokes were immediately exchanged over the police radio and with the repairman. Neither Spaur nor Neff took the subject seriously.

The deputies then headed west on Route 224 with the intention of filling out an accident report at the hospital. They saw a car parked on the shoulder on the south side of the road. They turned their patrol car around and approached the abandoned car from the rear. Spaur reported what happened:

> He [Neff] gets out the right side, I got out the left side, he goes to the right front corner of the cruiser, which is where he stops—sort of an insurance policy—and I went to the left rear of the other vehicle. I turned just to make a sort of visual observation of the area, to make sure nobody had walked into the woods, you know, to take a leak or something. And I always look behind me so no one can come up behind me. And when I looked in this wooded area behind us, I saw this thing. At this time it was coming up. And there's a slight rise there; went up to about treetop level, I'd say about a hundred feet. It started moving toward us—well, now, the trees that it was clearing

> were right on top of this rise right beside the road. . . . And at the time I was watching it. It was so low that you couldn't see it until it was right on top of you. I looked at Barney [Neff], and he was still watching the car, the car in front of us—and the thing kept getting brighter and brighter and the area started to get light, and I looked at Barney this time and then told him to look over his shoulder. So he did. He didn't say nothing, he just stood there with his mouth open for a minute, and as bright as it was. he looked down. And I started looking down. I looked at my hands, and my clothes weren't burning or anything when it stopped, right over on top of us. The only thing, the only sound in the whole area was a hum. It wasn't anything screaming or real wild. And it'd change a little bit—it'd sound like a transformer being loaded or an overloaded transformer when it changed.
>
> I was pretty scared for a couple of minutes; as a matter of fact, I was petrified; so I moved my right foot, and everything seemed to work all right. And evidently he made the same decision I did, to get something between me and it. So we both went for the car, we got in the car, and we set there. I wouldn't even venture if it was ten seconds, thirty seconds, or three minutes—and it stood there, and it hovered, and we didn't make any—anything—and it moved right out east of us [they were now facing east] and sat there for a second, and nothing still didn't happen to me, and Barney looked all right. I punched the mike button, and the light came on, so I picked it up. I first started to tell them, you know, this thing was there. And I thought, well, if I do, he'll think—so I just told Bob on the radio, I said, "This bright object is right here, the one that everybody says is going over." And he comes back with, "Shoot it!" This thing was, uh, no toy; this—hell, it was big as a house! And it was very bright; it'd make your eyes water.

They were ordered to follow the apparition, and thus began, perhaps, the wildest UFO chase on record. For more than seventy miles the object was chased, at speeds sometimes as high as 105 miles per hour.

While the chase was in progress, Officer Wayne Houston, in his police cruiser near East Palestine, Ohio, some forty miles to the east of the starting point of the chase, was monitoring the radio conversation between Spaur and his office in Ravenna.

Later, in signed testimony, Huston admitted to Weitzel:

> I talked with Spaur by radio. I met him at the north edge of the city on Route 14. I saw the thing when Dale was about five miles away from me. It was running down Route 14 about 800-900 feet up when it came by. This was the lowest I ever saw it.
>
> As it flew by, I was standing by my cruiser. I watched it go right overhead. It was shaped something like an ice cream cone with a sort of partly melted down top. The point part of the cone was underneath; the top was sort of like a dome. Spaur and Neff came down the road right after it. I fell in behind them. We were going eighty to eighty-five miles an hour, a couple of times around 105 miles an hour. At one point at least I was almost on Spaur's bumper, and we checked with each other what we saw. It was right straight ahead of us, a half to three quarters of a mile ahead.
>
> I am familiar enough with Rochester [they were now in Pennsylvania, some fifteen miles east of the Ohio border], and I guided him by radio. All the way we were trying to get contact with a Pennsylvania car. Had the base call Chippewa State Police station to see if they had a car on 51; they didn't. The first Pennsylvania car we saw was in Conway [a few miles east of Rochester]. Dale was low on gas, and we stopped where Frank Panzanella was parked.

Thus there enters the fourth observer: Frank Panzanella, police officer in Conway. His signed testimony reads:

> At 5:20 A.M. stopped at Conway Hotel and had a cup of coffee. I then left the hotel coming down Second Avenue. Looked to my right and saw a shining object. I thought it was a reflection off a plane. I then got out of the police car and looked at the object again. I saw two other patrol cars pull up, and the officers got out and asked me if I saw it. They pointed to the object, and I told them I had been watching it for the last ten minutes. The object was the shape of half of a football, was very bright and about twenty-five to thirty-five feet in diameter. The object then moved out toward Harmony Township approximately at 1,000 feet high; then it stopped and went straight up real fast to about 3,500 feet [and, according to other testimony, stopped]. I then called the base station and told the radio operator to

> notify the Pittsburgh airport. He asked me if I was sick. I told him if I was sick, so were the other three patrolmen. The object continued to go upward until it got as small as a ballpoint pen. Relative to the moon, the object was quite distant and to the left of the moon [Venus was to the right of the moon]. I could not see the moon from my position. The object was seen between two antennas in the backyard across the street to the east. We all four watched the object shoot straight up and disappear.

The object was hovering when the plane taking off from the airport passed under it, then took off directly upward, according to all witnesses.

Major Quintanilla, then head of Project Blue Book, attempted to establish the interpretation that all four police officers, who were sequentially and independently involved, had first seen a satellite (even though no satellite was visible at that time over Ohio[4]) and somehow had transferred their attention to Venus (which was seen by the observers while the object was also in sight). The original "investigation" was perfunctory; the initial inquiry, made of only one witness, Spaur, was a two and one-half minute phone call, which, according to Spaur, began with the words, "Tell me about this mirage you saw." The second interview, also by phone, lasted only one and one-half minutes. According to a signed statement by Spaur, Quintanilla apparently wanted Spaur to say he had seen the UFO for only a few minutes; when told that it had been in sight almost continuously while the observers chased it from Ohio into Pennsylvania, a distance of some 60 miles, he quickly terminated the discussion.

Quintanilla's method was simple: disregard any evidence that was counter to his hypothesis. Less than five minutes of phone interview sufficed for Blue Book to come to a "solution" of the case; only after Congressional pressure did Quintanilla travel to Ravenna, Ohio, to the Portage County sheriff's office, to interview Spaur and Barney Neff.

The interview was taped by Weitzel at the request of Spaur, and it provides a rare insight into Project Blue Book. This time the interview was long and involved. In addition to testimony from Spaur and Neff, it included testimony from Deputy Sheriff Robert Wilson, the radio operator who had been in radio contact with Spaur and Neff, and Sheriff Ross Dustman, whose chief role was to vouch for the character of his deputy sheriffs. However, it excluded two prime witnesses, Patrolman Huston of East Palestine, Ohio, who joined Spaur and Neff in the chase after their car arrived in East Palestine, and Patrolman Panzanella, of Conway, Pennsylvania, who joined all three in the sighting when the chase reached his town.

Because of the length of the taped interview only excerpts can be given, and these of necessity will be out of context.[5]

SPAUR: Second of all, I'm under the impression that Venus rises out of the east, as the morning star. And this is probably another thing that's wrong, I'm not sure.

QUINTANILLA: Depends, depends.

S: Huh?

Q: Sometimes it'll rise right over you.

S: Oh. O.K. So anyway . . .

Q: Venus, Venus—Venus. today [papers rattling] rises at 2:49 in the morning. And it rises 150° azimuth and 25° elevation. It doesn't have to rise low on the horizon; it can rise high. But it's on the ecliptic, yes.

S: O.K., so it's on the ecliptic. Granted you have this. Now this, this thing is this large, this big, and this low, and these people watched this thing from over in the Mogadore area; they report it, and I follow it, and I have Barney with me. We're going down the road; so you're gonna discount, well, there's two nuts; we're running Venus. Now Venus . . .

Q: Now, wait a minute . . .

S: Well, wait a minute, let me speak . . .

Q: You used the wrong word . . .

S: O.K. Well . . .

Q: I'm an officer in the United States Air Force . . .

S: Right. You definitely are . . .

Q: And I don't call anybody a nut.

S: No, O.K. I have hallucinations then! But this is what I've been saying . . .

Q: I didn't say you were having hallucinations.

S: What I'm trying to say is this. I'm going down the road; now this thing that I am following . . .

Q: And treat me with the same respect that I treat you.

S: I will sir; I am. I'll treat you with more respect than I've been treated the last . . .

Q: I'm not calling you a nut. I'm not saying you had hallucinations.

S: All right, the last twenty days! Anyway, this thing passes over another police car. He watches it go by; he's spotted it now. This is two cars that are fixed on Venus. So we're going down the road. And we get into Conway, Pennsylvania, and then this thing passes over the third car that's sitting there. Not even on the same frequency [a reference to the fact that he and this patrolman could not have been in communication prior to the event]. I never met, seen, spoke to before nor after this another officer. He's watching the same thing as it goes over top of him, going toward Pittsburgh, as we come screaming in. Now: we watched it, four men, standing right there, four officers. Probably you say anything you want, we stood right there, watched it, watched the plane go underneath it [a reference to a plane that had just taken off from the Pittsburgh airport], and we watched it make a vertical climb straight up. And this, sir . . .

Q: Disappeared.

S: My knowledge is God's truth. Yes, sir. The only thing left even to look at, after we went to the station and called the guy [the radio operator had relayed a message to call "a colonel or something"], was the one bright spot that was there. The sun was coming up full, and the moon was fading out. It was about a quarter of a moon [actually just four days before new moon], and right straight off that moon, which would have been to the south of the moon if you were looking west [east?], was one bright spot. I'd say it was probably, would look like a pencil eraser, real bright. [This of course was Venus, yet Quintanilla still stuck stubbornly to the Venus hypothesis for the sighting.]

WILSON: [the radio operator who had monitored the radio but had not viewed the UFO]: That was the mother ship.

S: Huh? The other ship?

W: That was the mother ship!

S: Oh, the mother ship. You guys are gonna have me convinced pretty soon. Aw, give me a tranquilizer and some coffee. . . [laughter] This thing was to, would have been to the left, which was the north of it, and we watched it, and it went up, stopped, the airliner went under it, and then it went straight up. Just as straight up as, well, just straight up. And there—I, uh, I wouldn't conceive of what, I know people can get fixed on something maybe, or something like that; but I don't think that—I don't see how myself and another cruiser and another guy and all this could go over. Chasing Venus. I, uh, I won't concede a part of it. I know that there's—this may be a way to discount it or what it is, but I know it was there. I seen it very plainly.

Q: Dale, it's not a question of discounting; we're trying to get into the [one word fuzzy]. We're trying to make the determination as to what it was.

S: Sir, if I could tell you what it was, believe me, Major, I—I myself—and like I said before, if I told you that I seen a Ford going down the highway, you'd know what I was talking about. And if you said, "Gee, there goes a Chevrolet," you would assume the fact that you identified it, and I would know what it is. The same thing with an aircraft. You say, "There goes a B-29," and I say, "Yep, sure is, that's an old war horse," or something to this effect, and it's identified. This, I have never seen nothing like it before or after or in the wildest farfetched imagination. I know you can have an optical illusion or even see something moving or like if you look through a piece of glass or something . . .

Q: Yeah, distortions.

S: I can go along with this. But nothing this big. In my wildest dreams I don't think I could have ever imagined or seen anything like it. But this thing was there. I seen it very plainly; I seen it outside the car. I saw it inside the car, and I saw it from outside the car after I got to Conway. And I would hate to think that I gambled this man's life [Neff] and a lot of other people's lives chasing Venus. I don't believe for an instant that I was following Venus. I don't know how to explain it. I don't have the slightest idea. But sir, this thing was as real as [indistinct word] . . .

Q: You know, Dale, I'm just going to say this for whatever it's worth: you're not the first one it's happened to.

W: [radio operator]: What does the Air Force think these are, Major?

Q: Misinterpretations of conventional objects and natural phenomena. Last year we had 245 astronomical cases.

W: What category does this go under, what Dale saw?

Q: Place it in the category of satellite and astronomical observations.

This case now appears in Blue Book statistics as an observation of Venus even though the object *and* Venus were reported to have been seen.

Four different sets of human eyes reported something to their respective brains, four brains that were accustomed to making evaluations of what their eyes observed. Two observers were in one car; each of the others was in separate towns. The testimony of the other two policemen was never obtained.

Quintanilla was obviously satisfied that the requirements of the scientific method had been met. He would, indeed, have been satisfied with four minutes of testimony over the phone had not Congressman Stanton, who had taken a personal interest in this case, forced his hand.

I have devoted considerable space to this incident because it is representative of my experience with Blue Book over many years as consultant. What I considered obvious cases of misinterpretation and unreliable reporting Blue Book would take some pains to establish for the record; cases such as this, which were open to question and contained the possibility that something "genuinely new and empirical" might be contained in it, were treated with little or no interest.

Had the observers in this case not been police officers, I feel certain the evaluation would have been "unreliable witnesses," a favorite category for cases in which the witness could not defend himself. To call a policeman an unreliable witness would clearly not have been politic, so the virtually untenable category of "astronomical" was chosen even against the advice of the astronomical consultant.

It should be apparent to any discerning reader that two issues are interwoven in this entire matter: one is the question of the reality of the reported UFO phenomena; the other is the matter of scientific methodology and scientific integrity. Regardless of how the first issue is resolved in time, the record will show that once again in the long history of science prejudice, emotion, and temporal provincialism marred, in the case of UFO research, the otherwise largely exemplary march of science and intellectual adventure.

The Portage County case was especially embarrassing to me since it had been repeatedly stated that Blue Book adopted no astronomical interpretation of a UFO sighting without my concurrence as consultant astronomer, but the rule was frequently and flagrantly violated. In this instance, the evaluation of this case as "satellite and Venus" was made without any consultation with me.

Three months later I was sent the Blue Book file on the case; my evaluation was a strong Unidentified—an evaluation that was strongly supported by the fact that it had been established by means of taped testimony that the observers had seen Venus as *well* as the UFO. The officers did not know Venus by name, but they confirmed that there was a "bright spot near the moon." On that morning Venus was just a few degrees to the upper right of the moon. The observers indicated that as the dawn light increased just before sunrise, the silhouette of the UFO became more distinguishable; quite the opposite would have happened with Venus as dawn light brightened. The sun rose that day at 5:42 A.M., and the sighting was terminated shortly after that. It didn't matter. My advice was not taken.

I have presented aspects of this case in some detail because although it is just one of a great many similar cases, it is a fine example in one instance of a Close Encounter of the First Kind, of the unimaginative attitude of the establishment, and of the "'real" nature of the experience for the observer.

The sequel to this case is not pleasant. Largely because the press and Blue Book concentrated on Dale Spaur almost to the exclusion of the other three witnesses, the public gained the impression that here was a case of one policeman's having become unbalanced and having experienced a major hallucination. It is clear that this certainly is the implication in Quintanilla's interview with Spaur. Subsequently, Spaur was singled out for unbearable ridicule and the pressure of unfavorable publicity. The combination of events wrecked his home life, estranged him from his wife, and ruined his career and his health. He is no longer with the police force, and, it is reported, he subsists by doing odd jobs.

Tragic denouements are fortunately not a part of the prototype of Close Encounters of the First Kind. But the Portage County case and the others chosen as representative in this chapter do portray the nature of the UFO when reportedly experienced close at hand.

Brilliant luminescence, relatively small size (of the order of tens rather than hundreds of feet), generally oval shape—sometimes capped with a dome—absence of conventional wings, wheels, or other protuberances, and ability to hover and to accelerate very rapidly to high speeds characterize the UFO at close encounter. Localization of appearance is likewise a salient characteristic. UFO trajectories are largely vertical when speeds are high—takeoffs at forty-five degrees or greater seem to be the rule. There is little tendency for the UFO to "cruise about the country" except locally.

So far in this category of Close Encounters the UFO has not left its mark except on the memories of the percipients. Now we turn to Close Encounters that do leave their marks—on inanimate or animate matter. Because marks can be measured and studied, therein lies their special importance for scientific investigation.

CHAPTER NINE

Close Encounters of the Second Kind

I assumed as a matter of course that it was a totally new invention and fervently hoped that the inventors were our own people.

—FROM A PERSONAL REPORT TO THE AUTHOR BY AN ARMY CAPTAIN STATIONED IN OKINAWA. THE SIGHTING WAS MADE IN AUGUST, 1945.

When the reported UFO, generally a brightly illuminated "craft," leaves a visible record of its visit or encounter with human observers, this constitutes a Close Encounter of the Second Kind. Other than the fact that a physical effect of some sort is left as a memento, this category does not seem to differ in many ways from Close Encounters of the First Kind. Why in one instance the encounter is without physical incident while in the other a measurable physical effect on either animate or inanimate matter is manifested is a puzzle.

The physical effects reportedly include tangible marks on the ground that can remain in evidence for days or even months and come ostensibly from physical contact of the craft with the ground, the scorching or blighting of growing things (particularly plants and trees), discomfort to animals as evidenced by their behavior, and such physical effects on the human observer as temporary paralysis, numbness, a feeling of heat, and other discomfort. Interference with the local gravitational field sometimes is also reported, as evidenced by the reports of some observers of temporary feelings of weightlessness or other inertial effects, as though the well-known laws of inertia had been temporarily abrogated.

One remarkable reported physical effect involves interference in electrical circuits, causing car engines to cease functioning temporarily, radios to cut out or to exhibit uncommon static, car headlights to dim or be extinguished for a short while, and, on occasion, car batteries to overheat and deteriorate rapidly.

The significance of such physical interactions is obvious; they offer opportunity for physical measurement and the promise of hard data. Yet the treatment of such reports as old wives' tales or as the product of deranged minds or hoaxes has most unfortunately led to the almost complete absence of serious investigation and to the subsequent loss of the very hard data so tantalizingly accessible.

Despite the bizarre nature of the reports and the seeming impossibility of their having happened, the fundamental question is, as before, not *could* these reported things have happened but *did* they happen, more or less as reported.

I would not be engaged in delineating these matters in this book had not the evidence I personally have examined over the past years seemed overwhelmingly to indicate yes as the answer to the latter question. The bizarre events actually did occur, as unthinkable as this may seem to the physical scientist.

The introduction of tangible physical effects that do not seem to suggest mass hysteria and hallucination or even the psychic and the occult (unless we deal here with a form of poltergeist phenomena) introduces a new dimension in the study. My opinion may count little with my peers, but this is precisely why much greater depth investigation of such cases is necessary, to establish to the satisfaction of the physicist, in particular, that the reported events did in fact occur.

At present the average physicist dismisses the entire phenomenon as impossible. They are entirely correct to do so, in *their* frame of reference, for from the standpoint of our present knowledge of the way nature works, "such things just can't happen." But "stones couldn't fall from the sky," either, and "ball lightning is sheer nonsense." The story of the self-assured but untutored man visiting the zoo for the first time comes to mind. Upon seeing the giraffe he turned away with remark, "There ain't no such animal." So, of course, there are no such things as physical effects from UFOs. We have tangible proof of a giraffe; do we in Close Encounters of the Second Kind have tangible proof of UFOs?

The reader at this point may well interject, "But if these physical effects happen, where are the photographs of them, where are the plaster casts of landing marks, where are the fully documented accounts of car stoppages?" That is precisely the point. When the subject is greeted with such utter disdain as the UFO has been, the very obtaining of such data is immeasurably difficult. Without funds, without time, and often without the cooperation of the original observers, who fear ridicule by involvement, the kind of documentation needed in the court of science is virtually unobtainable. To secure it one must travel, one must telephone, one must work at top speed. Above all one needs time, and it would be helpful but not necessary to have the sympathetic understanding of one's colleagues in engaging in such work.

In connection with the reliability of Second Kind sightings, it is interesting to note that if we refer to *all* cases of landing marks, regardless of the number of witnesses, the catalog developed by Ted Phillips contains cases from twenty-four different countries, the six leading countries being the United States, Canada, France, Australia, Spain, and Argentina. Since this happens also to be (with the exception of England) essentially the lineup of countries in which UFO investigation is the most active it probably follows that the phenomenon is truly worldwide.

My experience in the investigation of UFO Close Encounters of the Second Kind once again convinced me that the ubiquitous "real experience" phenomenon is present. There is no doubt that to the reporter of the event the experience was real—traumatically real in some instances. What is to the point, the physical effects—the semipermanent ground markings, for example—effects that could be photographed, also were real.

For this reason Close Encounters of the Second Kind bear a special importance, for when it is reported that a UFO has left tangible evidence of its presence, here is clearly the area in which to begin digging for scientific paydirt. Here is where new investigative efforts offer the greatest promise. It is in this category of UFO reports that we find the real challenge to scientific inquiry.

In the cases of Close Encounters of the Second Kind used in this chapter, the usual standards prevail. Only cases with multiple witnesses are used, although there exist very striking examples that had only one observer. The average number of observers in these selected cases is 4.0; the median, 3.0. I have included nearly twice the number of cases used in each of the previous categories because of the different types of physical effects reported so that we can examine several cases of each main type of physical effect (automobile stoppages, marks on the ground, etc.).

There seems to be a significant shift in the occupations of the observers in this category as compared to those of the earlier groups, which had a larger share of pilots, officers, and well trained technical persons.[1] Housewives, teenagers, and businessmen predominate in Close Encounters of the Second Kind. In this category, therefore, let us see what *combinations* of observers occurred. The case designation, the observer combination, and a very brief statement of circumstances are included in Table 1.

Table 1

OBSERVER COMBINATIONS IN SELECTED CLOSE ENCOUNTER (SECOND KIND) CASES

CEII-1	Six adult males, various occupations, and two teenagers (one a college freshman). All independently had a similar experience within two hours within a rectangular Texas area about thirty by twenty miles. Late night, misty, and always on open, lonely road.
CEII-2	Schoolteacher and ten-year-old son. Lonely road near small Wisconsin town. Night.
CEII-3	Chief of technical service, Air France; three pilots and three engineers. Tananarive, Madagascar. Early evening.
CEII-4	Supervisor in mailorder house and collection manager, finance company. Near bridge on lonely road. Night.
CEII-5	Husband and wife, painter and hairdresser, respectively. Out driving at 1:00 A.M. to see snow cover from recent storm. Passing cemetery.
CEII-6	Two businessmen traveling in separate cars. Road outside Virginia town, 8:40 A.M.
CEII-7	Nineteen-year-old roofer, father (forty-six), and grandfather (seventy-two), farmers. 4:00 A.M. on farm.

CEII-8	Three teenaged females, high school students. One editor-in-chief of yearbook, cheerleader, and officer in various clubs. The second (driver of the car), member of National Honor Society, editor-in-chief of school paper, majorette, French and College clubs, member of Math and Physics Club. The third, member of Honor Society, majorette, and member of several school organizations. Outskirts of town, lonely area bordered by woods. Night.
CEII-9	Three teenaged males, one teenaged female. Dusk. Teenagers were milking cows on farm.
CEII-10	Engineer, wife, and small son. Driving on lonely road in Oklahoma. Weather misty, low cloud ceiling. Dusk.
CEII-11	Two police officers. Eleven P.M. Open road in Texas.
CEII-12	Farmer, teenaged daughter, and teenaged girl cousin. Late night. Farm in Iowa.
CEII-13	Two businessmen and their wives. Late night on open country road.
CEII-14	Professional artist and husband. Night. Small town in Kentucky.
CEII-15	Adult male and wife. Road in Florida. Late afternoon.
CEII-16	Two elderly women and, independently, a beekeeper. France.
CEII-17	Businessman, wife, and their three teenaged daughters. Small town in Wisconsin. Night.
CEII-18	Two adult males, employees of a Canadian tourist fishing resort, their wives, and members of families. On lake. Late at night.
CEII-19	Nine teenagers (five girls, four boys), four housewives, and one adult male. Shore of lake in upper Michigan peninsula.
CEII-20	Senior state highway designer, wife, and mother-in law. Highway in open country. Ten-thirty P.M.
CEII-21	Woman and her three teenaged daughters. Small town in state of Washington. Night.
CEII-22	Cowboy and friend.

The isolation of the observers at the time of the sightings and the presence of highly educated or trained people in only three or four of the twenty-two cases seem significant. Does this make the reported events less credible, or is it possible that more sophisticated individuals refrain from reporting such "unbelievable" events? According to the system used in this book, it is clearly necessary to assign a lower Probability Rating to these cases. Yet interrogations revealed no less sincere amazement and puzzlement and no less a sense of having had a "real experience" than was the case among the more highly trained observers found in the categories already examined.

From my own considerable interrogation of witnesses as well as from the many accounts from tape recordings made by other investigators well known to me, I can testify that, in particular, Close Encounters of the Second Kind impressed the observers with a sense of vivid reality.

Pages here could be taken up with accounts from witnesses in near-hysteria as they told their story to police officers and others (generally not to me because in my Blue Book investigations I often arrived on the scene many days after the event); of physiological and psychological after-effects (there is no evidence that the cart is before the horse; the hysteria and the psychological disturbances came after, not before, the event); of disturbed dreams for weeks thereafter and sometimes even of changed life outlook and philosophy stemming from the encounter. To a few it has been akin to a religious experience, but since several witnesses were generally involved (whereas religious experiences are intensely personal events), their experiences cannot be so classified.

The physical proximity of the event would certainly tend to make the experience vivid and unforgettable. In one case the car in which four persons were riding was mysteriously stopped, and the lights and radio became inoperable during the short interlude during which a brilliantly lighted object hovered just ahead of the car. The policeman (see Appendix 1, CEII-13) to whom the report was later made stated: "All four of the people in the car appeared to be badly scared. The driver of the car did most of the talking. The two men were in the front seat, and the women were in the back." The other male was said to be in such a state that he "just couldn't make his words come out." It was reported that his voice quivered and that he trembled noticeably.

In the classic Loch Raven, Delaware, case the car in which two men were driving was involuntarily stopped as they approached a bridge over which there hovered a brilliantly lighted UFO. (See Appendix 1, CEII-4.) The men stated in an Air Force interview: "Then we decided to put the car between ourselves and the object. It was a very narrow road: on one side was the lake and on the other, a cliff. There was no place to run. We probably would have if we could, but we were terrified at what we saw."

The witnesses usually try to rationalize the event to themselves, almost invariably becoming frustrated, and I am personally convinced that many people recounted their experiences solely because they wanted desperately to know whether anyone else shared the same or a similar experience. Many have told me that were they ever to have another such experience, they would *never* report it.

I know from personal contact with many airline pilots that under no circumstances would they officially report their experiences. They know better. Some have informed me that they wished to forget that the whole thing ever happened. With such people it has been only after the greatest persuasion and upon my word of honor that their accounts and names would never be publicly used that I have been able to obtain their stories.[2]

To turn to the physical effects reported in this category of sightings, perhaps the most intriguing—and certainly one of the most difficult to explain in terms of our present knowledge of the physical world—are the globally reported cases in which a UFO is said to have interfered with moving automobiles by killing a car engine, extinguishing the lights, etc.

Why *this* physical effect, of all things? There would seem to be so many other, more significant ways in which UFOs could interfere in human affairs! Yet this is what is reported: cars are seemingly accosted on lonely roads, sometimes but not always resulting in a killed engine and the failure of lights and radio. It would almost seem as if the UFO regarded the cars as creatures to be investigated. This is the impression one gets from interrogation of observers and from a study of their reports. But ours is not to ask why (at least not until we have more facts); we examine what has been reported, choosing reports given by who seem to be the most credible witnesses.

We can start building the prototype of this subset of cases with one that reportedly occurred on a lonely road outside a small town in Wisconsin at night in early spring. (See Appendix 1, CEII-2.) I start with this because during my interview with her the principal witness (schoolteacher and former Air Force flight attendant) quite incidentally gave physical testimony—a description of feeling momentarily weightless—that might conceivably furnish a clue to the nature of the phenomenon.

The witness described the event thus:

> . . . that thing came from the dip in the hill, real fast but real, real smooth like something gliding, but lower than any plane, and hovered

> and stopped above that car [a car that had just previously passed the observer's car]. Then is when its [the other car's] lights went out, and I pulled onto the gravel because I thought it was a kid. He put out his lights, and I didn't want to smash into him—at all of this my lights were dimming slightly, but I didn't think anything of it until my engine, lights, and radio went out and stopped. This happened to me when it [the UFO] left that car and came down the highway . . . and was above us. It came down over from the other car. It was pretty low. When I looked out of my windshield I had to bend forward toward the wheel, and I looked straight up and there it was above us—with the car dead. I had opened the window when the other car's lights went out, and it was open then—and absolutely no sound.

QUESTION: Were you conscious at all of stopping your car, or did the motor go out entirely by itself?

ANSWER: No, I stopped it.

Q: You stopped the car?

A: and the car was running

Q: Well, I mean the engine.

A: Yes, the engine was still running.

Q: And then what happened?

A: . . . and then this red object came, it hovered, it came above us. And all of a sudden everything got real still . . .

Q: Well, now, tell me this. If you had some magic way of putting something up in the sky that closely resembled what you saw, what more or less common thing that you have around would you put up there that would most closely resemble in shape what you saw?

A: Well, you know those rolls that you buy of Bisquick or Pillsbury and they're in that little tube in your refrigerator case in the store, and you rap on the side of your counter and then you get out triangle shaped dough and then you roll that up and it looks like a crescent shape? That's what it would look like.

Q: I see. Well, I'm not exactly a cook, but I can figure this. Let's see, are you acquainted with Australian Boomerangs?

A: Australian?

Q: Well, boomerangs. You know what a boomerang looks like?

A: I've never had one. It would be like that, except it was more rolled than flat.

Q: Now, you kept calling it a red color. What shade of red?

A: Oil paint. The best color I can say it would be is an orangeish-red. . . . And it was like an Indian sunset or something in color.

Q: Did it appear to be a solid object, or did it appear to be mostly light?

A: Well, when it came above us, then it was definite. I mean there was a definite pattern, but it seemed to be more solid, and then toward the edges it was more like [fuzzy].

Q: Did it ever stand still?

A: Uh huh, when it was in the air it did. Of course, it was always in the air, but when it stood right above us [it was], and I tried to start the car and I tried and I tried, and as long as that thing was above us, I just couldn't get that car to go. It just didn't even want to—it just nothing. It wouldn't even turn over, just grunt a little bit and that was it [at this point witness gave a graphic description, with sound effects, of the futile noises the starter made as she tried desperately to start the car]. Swell, I turned the key and it went ugh, and that was all. Then it didn't do anything. It was like a dead battery.

Q: Well, now, when it left, did it go up or sideways or what?

A: No, it didn't go straight up. It went behind us on my side, and it went over in the field toward a farmhouse there. . . . It just went real smooth, and it didn't hesitate, and it didn't jerk.

Q: How long did it take to disappear?

A: It didn't right away. Finally when it left the car [it] sort of jerked. I turned [the key] over, and it went *ur-ur-ur*, and then finally it turned

> over real good, and I finally got the car started. . . . By that time I had floored the car, and I had gotten up to Cochrane by the mill there. . . . And I saw it crossing the railroad tracks, and it was going slowly down [the tracks].

The interview with the schoolteacher was lengthy. One other fragment not only makes the feelings of the reporter clear but also describes a phenomenon, reported in other cases also, which may point to the physics of the UFO:

> . . . you know, if you stay in a house at night and everything is still, there are still the noises of the living, you know, but when this thing was there, there wasn't even the noise of living. It was nothing. It was an eerie quiet. . . . Another thing I remember . . . as though I was light in weight and airy. Something like the first time you experience an airplane takeoff or drop from an air pocket. It felt like the air and everything was light and weightless.
>
> One thing I remember—my feet burning for some time after. When I first stepped out of the car, it felt like scalding dry heat on them. I always thought if I saw one of these things I'd just get out and walk up to it, but it didn't give any inkling of being an earthly thing, so I just stayed in the car, which was completely dead, and I couldn't go any place. I guess I was just waiting for I don't know what.

Here now is a very brief synopsis of another report, which never would have been made had not an interested person overheard a remark at a basketball game made by people unknown to him. He made it a point to talk to them and made the initial report for them. They later consented to be interviewed by Raymond Fowler and his New England colleagues. It was a "typical" Close Encounter—starting first with the lighted craft, which they first took to be a helicopter, seen at some distance. It soon approached, as the car and the UFO traveled toward each other, and the car and its electrical system became inoperative.

Excerpts from the taped interview will give us the experience in their own words. (See Appendix 1, CEII-8.)

> Janice noticed the object, so Kim pulled over. They wanted to get out of the car, but I didn't. All of a sudden the car stalled, and the

> radio and the lights went off. Then nobody wanted to get out of the car. Truthfully, I was too scared to carefully observe the object. I just noticed the four lights when they passed. Kim finally got the car going.

Another witness to the same incident said:

> Janice said, "What's that?" I just glanced out of the window and said, "Must be a helicopter." Janice would not dismiss it as such, and then Kim became very excited. At first she [Kim] laughingly said, "It must be a UFO or flying saucer." All of a sudden it wasn't funny anymore. . . . I was scared, and I refused to get out of the car. We had just pulled the car over. Suddenly the car stalled and the radio and the lights went out. The object passed, and the car started.

Kim, the driver, said:

> When we got close to the object, the car stalled and the lights and our radio all went off at the same time. After this I tried to start the car twice while the object appeared to remain stationary. Thinking that the lights and radio would be drawing too much power from the battery I shut the light switch and the radio off. Then I tried to start the car again twice. It did not start. Next, the object in the sky seemed to start moving away from us. I tried to start the car again, and it immediately started, proving that it was not flooded. . . . Since we had replaced the battery in our car just three weeks ago [prior to the sighting], I do not believe it was the car's fault. I had the clutch in at all times since I was pulling off to the side of the road to stop.
>
> . . . I saw an object to the left of us in the sky, which at first appeared to be a plane. As we approached it, I saw that it was too large and too low to be a plane and called the attention of the other occupants to it. . . . The object was moving in the same direction as we were at first, then stopped for about a minute, then flew off, and the car started again. The object made no noise, and it did not affect the street lights in the area.[3]

The three highly intelligent witnesses were evidently plagued by the often encountered inability to put into practical, descriptive terms the elements of their sighting. For instance, in answer to the question about what they would place in the sky that would give the same appearance as that of the sighted object, Kim answered, "Erector set material with white lights reflecting on it with red lights on the top." Ellen's reply to the question was, "Four searchlights?" Janice stated, "The object was a regular trapezoid, although I could not make out its exact outline. There seemed to be a dim light on top, perhaps a small structure there."

"Red lights about as bright as a hot electric stove." "It was glowing around the white lights. It reflected like some sort of metal." "I have never seen anything like this before." "The object was too large to be any kind of aircraft. The shape was odd and did not resemble a balloon or helicopter at all." "The object hovered in a fixed position, then turned and disappeared in a westerly direction. It rose and flew out of sight." These fragments of interviews with several witnesses are hardly the sort of descriptions one might expect from honor students, editors of the school paper and its yearbook, had they been describing an ordinary aircraft, even if seen under unusual conditions.

In a case already referred to (see Appendix 1, CEII-13) in terms of witness reactions, the policeman who was the first to talk with the observers stated:

> While object was near, driver said car wouldn't accelerate—lost power and sputtered "like it wasn't getting enough juice." When object appeared closest to them, it was no longer bright but "a clear and well defined lens-shaped object with a dull light amber color—like a traffic caution light, only more pale in color." I don't think that anybody could possible have staged the facial expressions and fear that those people showed.

The sketch the driver later drew of the object reveals precious little detail, showing merely an egg-shaped object the surface of which was covered with inset objects resembling automobile headlights. "Each of these," the report states, "gave out a shaft of very bright white light, making the object as a whole appear to have rays of light extending outward in all directions. Later it looked like a well-defined lens-shaped object, amber in color."

The prototype of the Close Encounter of the Second Kind is further embellished by the account of a UFO sighting in a cemetery after midnight (see Appendix 1, CEII-5), of which one of the observers said, "Nothing I have ever seen compares with the object." The two reporters of this event had been riding late at night in the country purposely to look at the snow-laden branches of trees after a heavy snowstorm.

As they passed a cemetery, which seemed to be shrouded in fog despite an otherwise crystal clear night, a light shone in the midst of the fog. Thinking that there was a fire in the cemetery and that the fog was really smoke, they turned the car around after having gone a short distance and returned to the scene. The investigator's report reads:

> He turned the car around again and put his windows down and drove off the road broadside to the cemetery and to the light [which was directly over the cemetery]. . . . He got out of the car, shut the door [window open] and started to point to the object. Simultaneously several events occurred: the automobile lights, radio, and engine ceased functioning; he felt an electrical shock, and his body became numb and immobilized; the arm he was pointing with was pulled against the roof of the car and hit with such force that it left an imprint in the ice and snow Mr. W. could not move a muscle, although he could hear and his mind seemed to be functioning normally. Then the lights and radio came back on, and the object which had been rocking back and forth emitted a humming sound and accelerated upward and out of sight above the fog patch.

We have already referred to the Loch Raven Dam case in speaking of the reactions of witnesses. (See Appendix 1, CEII-4.) Now in terms of the object and its physical effects described we turn to a portion of a transcript from an Air Force interview of one of the witnesses:

> Shortly after you pass the dam . . . the bridge looms up in front of you at 200 to 250 yards away We saw from that distance what appeared to be a large, flat sort of egg-shaped object hanging between 100 to 150 feet off the top of the superstructure of the bridge over the lake.
>
> We slowed and then decided to go closer and investigate the object. . . . When we got to within eighty feet of the bridge, the car went completely dead on us. It seemed as though the electrical system was affected: the dash lights went out, the headlights went out, the motor went dead. Mr. S., who was driving the car, put on his brakes [after

> the motor went dead], turned the ignition once or twice. We didn't get any whirring sound; we were pretty frightened at this point. . . . We watched it . . . for approximately thirty to forty-five seconds and then, I am not sure of the sequence of events here, it seemed to flash a brilliant flash of white light, and we both felt heat on our faces. Concurrently there was a loud noise, which I interpreted as a dull explosion. . . . Then very quickly . . . the object started to rise vertically. It didn't change its position [aspect], as far as we could tell, during the rising. The only different feature it had while it was moving was that it was very bright and the edges became diffused so that we couldn't make out the shape as it rose. It took from five to ten seconds to disappear from view completely. We were very frightened. . . . We got back to a phone booth in approximately fifteen minutes. We proceeded to call the Ground Observer Corps, with no result. Our story elicited only complete disbelief.

Until the subject of UFOs has gained sufficient scientific respectability so that younger people with scientific imagination and courage can undertake proper investigations of the subject, we are left with most unsatisfactory descriptions of brilliantly illuminated oval objects that perform the most incredible feats. We shall have to content ourselves with saying that Close Encounters of the Second Kind involve a UFO that seems to have the strange property of being able, in some unfathomable way, to interfere with car ignitions.

How this could happen—as we must assume it does unless all the seemingly solid witnesses are pathological liars—is as foreign to our physics of 1972 as the origin of solar energy was to the physics of 1912. We knew then that the sun had sources of energy completely unknown to us; it was there and had been shining in the same way for hundreds of millions of years, as demonstrated by the fossil bones of animals that had lived hundreds of millions of years ago. But how it performed this trick of manufacturing energy seemingly out of nothing we did not know. In that case, however, we knew that it *did* happen; when our physics caught up with the sun, so to speak, we knew how it happened. In the matter of UFO Close Encounters with cars, we cannot yet *prove* beyond all doubt that what the observers reported really *did* happen. We are still in the stage of gathering data.

For the moment, let us look at the probability that motors are killed and lights and radio stop by coincidence when the driver has a UFO close sighting.

We have all seen cars stopped by the side of the road, hood up, waiting for tow trucks. It would be highly improbable that a car would become completely immobilized and then a few moments later "heal itself," yet it can happen. Perhaps, for example, a wire that had become loose was jarred back into place in some way. But to combine this low probability event with the simultaneous appearance of a strange light coming down from the sky and hovering over the car, the car remaining disabled only so long as the light was present, is dubious at best.

It is, of course, much the easier way out to dismiss the whole matter as "psychological" (whatever that means in this context) and return to commonplace, understandable matters. However, that would not be acting true to the high ideals of science, which involve being curious about all things that occur in man's environment, investigating and weighing them, and calmly considering the evidence.

If the probability of a happening in any *one* case is extremely low, consider the probability of coincidence in the following train of events—if they happened as reported.

On the evening of November 2, 1957, at about 11:00 P.M., just one hour after the Russians had launched their second, dog-carrying artificial satellite (that certainly was coincidence) but before we Americans knew about it, Patrolman A. J. Fowler, officer on duty at Levelland, Texas (population 10,000), received the first of several strangely similar phone calls. (See Appendix 1, CEII-1.)

The first was from Pedro Saucedo, who, with companion Joe Salaz, had been driving four miles west of Levelland when a torpedo-shaped, brilliantly illuminated object (as Saucedo described it) rapidly approached the car. Fowler listened to a terrified Saucedo relate the incredible story of how, as the object passed close over the car, the truck headlights went out, and the engine died. A certified copy of a statement made by Saucedo reads:

> To whom it may concern: on the date of November 2, 1957, I was traveling north and west on route 116, driving my truck. At about

> four miles out of Levelland, I saw a big flame, to my right front. . . . I thought it was lightning. But when this object had reached to my position it was different, because it put my truck motor out and lights. Then I stop, got out, and took a look, but it was so rapid and quite some heat that I had to hit the ground. It also had colors—yellow, white—and it looked like a torpedo, about 200 feet long, moving at about 600 to 800 miles an hour.

As the UFO moved into the distance, the truck lights reportedly came on by themselves, and Saucedo found that his truck started easily. The two men drove on to Whiteface, ten miles west of Levelland, and it was from a phone booth there that the call was made to Officer Fowler. Fowler apparently figured the man must have had one too many drinks, and he dismissed the report from his mind.

Considered by itself, the testimony of an uneducated, frightened truck driver, as sincere in his reporting as he might have been, has little credibility. But one hour later Fowler got another call, this time from Mr. W. of Whitharral. Fowler was told that he (Mr. W.) was driving four miles *east* of Levelland (the direction in which the Saucedo object had disappeared) when he came upon a brilliantly lit egg-shaped object, about 200 feet long, sitting in the middle of the road. As Mr. W. approached it, his car engine failed, and the headlights went out.

According to the observer, the object was lit up like a large neon light and cast a bright glare over the entire area. The observer decided to get out of his car, but when he did so, the UFO rose and, at an altitude of about 200 feet, the object's light or glare blinked out entirely. Mr. W. then had no trouble starting his car.

A short time later Officer Fowler got another call, from another Whitharral man, who was, at the time of the incident, some eleven miles north of Levelland. He reported to the police station that he had come across a glowing object sitting on the road and that as he approached it—the reader can finish the sentence—his car engine stopped, and his headlights went out. But when the object left shortly thereafter, all was again well.

But that was not the end. According to a signed statement in Project Blue Book files, at 12 :05 A.M. that Saturday night in November, a nineteen-year-old freshman from Texas Tech, driving roughly nine miles east of Levelland, found that his car engine began to sputter, the ammeter on the dash jumped to discharge then back to normal, and the motor "started cutting out like it was out of gas." The car rolled to a stop; then the headlights dimmed and several seconds later went out.

Baffled at the turn of events, he got out of his car and looked under the hood but found nothing wrong. Closing the hood, he turned away and then noticed for the first time, he reported, an oval-shaped object, flat on the bottom, sitting on the road ahead. He estimated it to be about 125 feet long, glowing with a bluish-green light. He stated that the object seemed to be made of an aluminum-like material, but no markings or other details were apparent. Frightened, he got back into the car and tried frantically but in vain to restart the car.

Resigned, he sat and watched the object sitting in front of him on the road (he did not state how close he thought he was to the object) for several minutes, hoping that another car would drive by. None did. The UFO finally rose into the air, "almost straight up," and disappeared "in a split instant." Afterward, the car was again fully operable.

"I then proceeded home very slowly," his statement continues, "and told no one of my sighting until my parents returned home from a weekend trip . . . for fear of public ridicule. They did convince me that I should report this, and I did so to the sheriff around 1:30 P.M. Sunday, November 3."

At 12:15 A.M. Officer Fowler got still another call, this from a man phoning from a booth near Whitharral. This observer reported his encounter with the strange object at a point some nine miles north of Levelland. Once again the glowing object was sitting on a dirt road, and as his car approached it, its lights went out and its motor stopped. Soon the object rose vertically, very swiftly, and when it reached an altitude of about 300 feet, *its* lights went off and it disappeared from sight. As the reader expects by now, at this point the car lights came back on and the car was started with no difficulty.

By this time Officer Fowler had finally realized that something odd was going on, and he notified the sheriff and his colleagues on duty, some of whom took to the roads to investigate. Two of them reported bright lights, seen for just a few seconds, but they did not have any car-stopping encounters.

At 12:45 A.M. another single witness—I have broken my rule to use only multiple-witness cases because of the independent witnessing of essentially the same event or object, with the same physical effects, from independent nearby points—driving just west of Levelland and thus close to the spot where two hours earlier Saucedo had had his sighting, spotted what looked like a big orange ball of fire at a distance of more than a mile. The ball then came closer and landed softly on the highway about a quarter of a mile ahead of the observer. It covered the paved portion of the highway.

The witness reported that the motor of the truck he was driving "conked out" and his headlights died. Meanwhile, the object sat there on the road ahead of him, glowing bright enough to light up the cab of his truck. In about a minute, the observer reported, it made a vertical ascent—and, of course, things returned to normal. This encounter was not phoned in at the time to Officer Fowler but was reported the following day. One possibly significant clue to some as yet unknown process may lie in the fact that the reporter stated that when the UFO landed it changed from its original red-orange color to a bluish green but that when it rose it changed back to red-orange. And it is perhaps of interest to note that the object or objects always landed on the pavement, except once, when it settled on a dirt road.

But that is not all. At 1:15 A.M. Officer Fowler got another call, this time from a terrified truck driver from Waco, Texas, who was at the time just northeast of Levelland, on the "Oklahoma flat road." The man told Fowler that his engine and headlights suddenly failed as he approached within 200 feet of a brilliant, glowing egg-shaped object. He said that it glowed intermittently "like a neon sign" and that he estimated it to be about 200 feet long. He reported that as he got out of the truck, the UFO quickly shot straight up with a roar and streaked away.

Officer Fowler stated that the truck driver was extremely excited when he called and that the witness was most upset by his close encounter. The truck engine and lights worked perfectly when the object left.

By this time patrol cars were out looking for the reported object. Sheriff Clem and Deputy Pat McCulloch were being kept up to date by Fowler as they drove around the area. At 1:30 A.M., while driving along the Oklahoma Flat Road, between four and five miles from Levelland, the two men spotted an oval-shaped light, "looking like a brilliant red sunset across the highway," a good 300 or 400 yards south of their patrol car. "It lit up the whole pavement in front of us for about two seconds," said Clem.

Patrolmen Lee Hargrove and Floyd Gavin were following in their patrol car several miles behind. In his signed statement Hargrove stated:

> Was driving south on the unmarked roadway known as the Oklahoma Flat Highway and was attempting to search for an unidentified object reported to the Levelland Police Department I saw a strange-looking flash, which looked to be down the roadway approximately a mile to a mile and a half. . . . The flash went from east to west and appeared to be close to the ground.

Constable Lloyd Ballen of Anton, Texas, also reported seeing the object, although his statement was: "It was traveling so fast that it appeared only as a flash of light moving from east to west."

None of these patrolmen's cars was affected, but Levelland Fire Marshal Ray Jones, who also was looking for the UFO, stated that his car's headlights dimmed and his engine sputtered but did not die, just as he spotted a "streak of light" north of the Oklahoma Flat.

Officer Fowler reported that a total of fifteen phone calls were made to the police station in direct reference to the UFO, and he added, "Everybody who called was very excited."

In terms of probabilities, that all seven cases of separate car disablement and subsequent rapid, automatic recovery after the passage of the strange illuminated craft, occurring within about two hours, could be attributed to coincidence is out of the statistical universe—if the reports are truly independent (and they are, according to the tests we've used throughout).

Suppose we try to attribute the happening to mass hysteria, although that does not disclose a mechanism for killing engines and extinguishing lights and stopping radios. The observers were independent unless all of them, for example, were listening to a local radio station that carried the news.[4] (No investigator ever checked into the important question of whether the radio stations were notified and if they broadcast the reports.) We know that at first Officer Fowler discounted the reports, and it is unlikely that he would have almost immediately notified the local station. But let us suppose that he or someone else did and that all car radios were tuned in to that particular station. We still would need an explanation for the physical effects reported unless we attribute them to downright prevarication rather than to hysteria.

What was needed at the time was swift reaction by Blue Book and a serious, thorough investigation. Captain Gregory, then head of Blue Book, did call me by phone, but at that time, as the person directly responsible for the tracking of the new Russian satellite, I was on a virtual around-the-clock duty and was unable to give it any attention whatever. I am not proud today that I hastily concurred in Captain Gregory's evaluation as "ball lightning" on the basis of information that an electrical storm had been in progress in the Levelland area at the time. That was shown not to be the case. Observers reported overcast and mist but no lightning. Besides, had I given it any thought whatever, I would soon have recognized the absence of any evidence that ball lightning can stop cars and put out headlights.

I was told that the Blue Book investigation consisted of the appearance of one man in civilian clothes at the sheriff's office at about 11:45 A.M. on November 5; he made two auto excursions during the day and then told Sheriff Clem that he was finished.

A newspaper reporter subsequently said that he had recognized the investigator and identified him as an Air Force sergeant.[5]

In any event, Blue Book came under severe pressure. In a memo dated December 4, 1957, Captain Gregory complained that ". . . as a result of pressure from both the press and public . . . Assistant Secretary of Defense requested that ATIC immediately submit a preliminary analysis to the press . . . a most difficult requirement in view of the limited data."[6]

Interfering with cars on the highways is but one of the physical effects reported in this category of Close Encounters. There are also the reported—and photographable—effects on living things, notably plants and trees.[7] Many witnesses have reported temporary paralysis in their limbs when their encounters have been quite close.

More than 300 cases of "scorched, denuded circles" and related "landing marks" frequently associated with the sighting of UFOs at close range have been cataloged. These, like UFOs in general, have been reported from many parts of the world, and a definite pattern is evident. The prototype is clear from an examination of even a few cases. Typically, in these cases a UFO looking, in most respects, like those in the first category or those of the second already described is seen to have landed or to be hovering near the ground. After it has departed, the witness finds a circular marking on the ground—sometimes nearly a perfect circle—which the witness invariably claims was not there previously. Of the cases cataloged by Phillips so far, 65 percent have occurred at night. If one chooses to examine only multiple-witness cases from the Phillips catalog, in keeping with our general policy, one must discard two-thirds of the cases. Yet from the nearly 100 cases remaining, witnesses reported that in three-quarters, the UFO was seen on the ground, and in nearly a fifth of them, at treetop level. In nearly all the multiple-witness cases the UFO is seen at or near the site of the later discovered marking.

The witnesses in these selected cases included some technically trained persons—medical doctor, airline pilot, engineer, ship's captain, mine supervisor—as well as farmer, factory worker, priest, patrol guard, etc.

The markings on the ground are discovered almost immediately in the daytime cases and the following morning in the more frequent nighttime sightings. Natural curiosity draws the witnesses to the landing spot, and there they generally find a marking that fits a general pattern: either a circular patch, uniformly depressed, burned, or dehydrated, or a ring the overall diameter of which can be thirty feet or more but which itself is one to three feet in thickness (that is, the inner and outer diameters of the ring differ by that amount, while the ring itself may be quite large). The most frequently reported diameters are twenty-thirty feet. It is almost universally reported that the rings persist for weeks or months—sometimes years—and that the interior of the ring or sometimes the whole circle remains barren for a season or two.

The main problem with the UFO rings is to establish that there was, indeed, a connection between the appearance of a UFO and the marks on the ground or, sometimes, with the scorched or blighted tops of trees. As might be expected, the tendency has been to dismiss the rings and landing marks from the purview of science, attributing them to hoaxes or natural causes, thus leaving the burden of investigation to a handful of private investigators, such as Ted Phillips.

Care must be taken not to confuse these "Close Encounter Rings" with the so-called fairy rings, which are nothing more than fungus growths in which the fungus, starting from a central spot, spreads outward in an ever-widening ring. No fungus that I know of can produce burned, charred, or scorched leaves or can give the leaves the appearance of having been subjected to intense heat from above.

Returning to the general plan of illustrating the prototype by means of synopses of selected individual cases, we start with one that both Phillips and I personally investigated. This sighting occurred in Iowa in July, 1969. Two teenaged girls stated that they were exceedingly frightened late one evening when, looking out from their farmhouse bedroom window, they espied the "traditional" lighted craft gliding away from the farmhouse, accompanied by a jetlike sound. (See Appendix 1, CEII-12.) The father of one of the girls, a farmer, had just that day examined his soybean field in preparation for cultivation and had found it in good order.

Shortly after the UFO sighting there was a light rain, and early the next morning the farmer went out to check whether the rain had been severe enough to interfere with his planned cultivation. To his surprise he found a forty-foot devastated circle in midfield where none had been less than a half day before. He had no explanation for it. He had learned of the girls' experience but had promptly discounted it until he saw his soybean field. The place where the girls had seen their object was not inconsistent with the position of the destroyed ring of plants.

I visited the farm several weeks after the event and saw the circular patch for myself. The leaves of each plant were hanging wilted from the stalks as though they had been subjected to intense heat, but the stalks themselves were not broken or bent, and there were no marks of any kind in the soil. Everything appeared as though the heat or destroying agent was applied directly from above and at close range but without direct contact.

The object that may have been associated with the circle was reported by the girls to have been observed at close range from their window, then to have turned to the northwest (it came from the south; the girls were looking out a north window, and the field was to the south of the house about a mile away) and disappeared, leaving only an orange glow in the sky. According to their report, it was spinning counterclockwise and had the shape of a shallow inverted bowl with a curved bottom.

It appeared to be dull gray-black metallic color with a circular reddish-orange band of light about two-thirds of the way from the bottom to the top. It was the illumination from the orange light that disclosed the shape of the object. No protrusions were visible, and there were no individual lights—only the band of orange light. In size it was described as three or four times the diameter of the moon, and one of the girls thought it appeared as large as an automobile would have at the estimated distance.

Because of local publicity the farmer refused to let the girls be interviewed by me but was himself fully accommodating in showing me the circle and answering questions. He wished no further publicity, made no attempt to capitalize on the event, and left me with the feeling that if the whole thing had in some way been a hoax, it would be difficult to discover any possible reason for his choosing to destroy a portion of his field (and by what means?) in the absence of any desire for publicity or monetary gain.

It has often been reported but seldom carefully documented that immediately after a close UFO encounter the top branches of adjacent trees have been found broken and the leaves wilted.

Here again is a fertile field for investigation. One case of this sort was made available to me through the kindness of Dr. Peter Millman, of the National Research Council of Canada, although he did not personally investigate it. The investigation report came from the Department of National Defense. Here is an excerpt from the report of the sighting, which occurred near the shore of a lake in northern Ontario on June 18, 1967. (See Appendix 1, CEII-18.)

> While returning home by boat from visiting neighbors [the two witnesses] noticed a bright object hovering fifty feet above treetops approximately a quarter of a mile away. Turned boat toward object to observe more closely when object suddenly and at a great speed descended toward boat. Mr. G. made a very hasty retreat, using full power of 75-horse outboard motor to make shore and get out of boat. Object then returned to original hovering position. Boat was reentered, and attempt was made to return to persons just visited, but object again appeared to make rapid descent toward boat. Boat was immediately grounded on shore, and Mr. and Mrs. G. ran to home of [another] Mr. G. and awoke entire household. Object hovered for about ten-fifteen minutes and then rapidly disappeared to west-northwest. . . . No noise heard at any time . . . wind conditions calm, but Mr. G. stated that tops of trees moved very noticeably as object made both descents.

Quoting from the official government report:

> The object was described as oval in shape with a slight rise on top as though with a canopy. Color was shiny . . . metallic and glassy. No lights were visible except that tops of trees appeared to glow white when the object tilted toward the descent but appeared to rise horizontally and fly horizontally when it disappeared. Object was very clear to the naked eye with moon reflecting off it. Mr. G. [estimated] size approximately twenty-five to thirty feet across and approximately ten to fifteen feet at thickest point. As object disappeared it took on an orange tint. No noise was heard by any of the observers nor those occupying two cabins at less than a quarter mile distance from where object was allegedly hovering. One occupant . . . did observe that he was listening to his transistor radio at time, station CKRC on 630 KCS, when so much static and interference was heard that radio was shut off. He looked out window . . . thinking thunderstorm in area but noted clear sky. His radio was checked and found in satisfactory operating condition No alcohol was consumed by any of

> witnesses evening of sighting. Mr. G. has good eyesight, needing no glasses. Several samples of wilting leaf limbs brought to Winnipeg for analysis.

The last statement refers to the unexplained damage to the tops of trees observers felt was associated with the appearance of the UFO. The report read:

> Department of Forestry and Rural Development advise they are unable to provide explanation for cause of wilting on three different types of trees: i.e., birch, hazel, and chokecherry, examined in the area from which previous samples had been obtained. There is no evidence of blight or insects. Several trees are affected but not in any fixed pattern and mainly on tops of trees. Forestry states cause could be heat, although no other [normal] evidence would indicate this as a source.

Dr. Millman has dismissed the light as almost certainly having been the planet Venus, which was indeed setting in the northwest at that time. But here we run into the hub of the entire UFO problem. No one bothered to find out, in this case, in what direction the observers were originally looking, and no one bothered to ask whether they saw Venus *and* the light. Of course, one might well wonder how two people of otherwise demonstrated stability could hallucinate to the extent that they believed Venus made two rapid descents toward their boat, causing them to use full power on the boat to make a getaway. This is another question that should have been investigated more thoroughly.

Case upon case can be adduced to build the prototype, but this would be of little avail. There seems to be no basic difference in the appearance of the Close Encounter cases that produce physical effects and of those that do not; in both categories far more detailed information is needed.

Perhaps we should let three final case synopses suffice:

> Mrs. J. . . . was attracted to the window by what appeared to be landing lights of an airplane. The light was extremely bright and seemed to be coming directly into the yard. . . . Fearing the lights were the landing lights of a crashing airplane headed directly toward the house, she herded the three girls in haste out of the house and into the yard away from the approaching lights, which by then had blended into one huge brilliant and intense white light. The four frightened

> witnesses stood in the yard . . . watching as the light moved in low in a straight line toward the house, suddenly lifted several hundred feet, clearing a clump of evergreens bordering the yard and dipped down on the far side of the tall trees and touched the ground. The intense light illuminated the surrounding area, including the side of the house and the yard. . . . Object was also [reportedly] observed by a border patrol officer who had been alerted by radio. He was "buzzed" by apparently the same object, which was low enough so that the patrolman stopped his car, got out, and watched it move out of sight. The four witnesses [at one location] and the independent officer gave the same description of the object as being about thirty feet in diameter, slightly domed, silent, and of a very intense white light.
>
> The object remained grounded for several minutes, and then it ascended almost vertically in a burst of speed and disappeared toward the northeast. Where the object had grounded in the sixteen or so inches of snow there was a large circular imprint about ten to twelve feet in diameter, and the ground beneath the melted snow ring showed evidence of having been scorched. Oval-shaped tracks eight inches long and eight inches apart, in single file, were found leading from the landing site to a clump of evergreens, where they disappeared. A month later the circular area still showed signs of the "landing."

I had no personal connection with the above case (see Appendix 1, CEII-21), but I have included it because it fits the pattern and was investigated by a competent UFO investigator.[8]

Another case takes us to the province of Quebec, Canada, where on May 11, 1969 (see Appendix 1, CEII-22), M. Chaput, a pulp mill worker, was awakened on his ten-acre farm at 2 A.M. by the barking of his dog. Looking outside he saw, the report of a practiced Canadian UFO investigator states, an intense light source illuminating his field about 600 feet away. It was so bright that it lighted the surrounding area, even the house. He went outside and could see his shadow cast on the house. When interviewed, Chaput said he felt that the light could not have been more than fifteen feet above the ground. Then the light vanished, but he could hear a purring sound receding in the distance.

The next morning he went to the scene of the incident with one of his teenaged sons and a younger child and found not only a circular mark of the traditional type but "three circular depressions, equally spaced to form a triangle with a rectangular depression one to two inches in depth," near the midpoint of the base of the triangle. The investigator conjectures that the three evenly spaced depressions might have been caused by landing legs or pads with the rectangular depression, the result of a hatchway pressing down with extreme force.

Conjecture aside, the fact remains that here is another of now more than 300 cataloged cases of ground markings reportedly associated with very close approaches of UFOs. The problem they present is both provocative and frustrating—frustrating because to get anywhere with the problem, far more quantitative data are needed than we have at present, although the pattern emerging from these worldwide reports seems sufficiently clear in outline.

One final case, with which I also had no connection but which is included in the Blue Book files as "Hoax," apparently was routed to Blue Book by an air attaché in Paris who, in turn, was apparently moved by a letter addressed to him through the Assistant Chief of Staff, Intelligence, Headquarters, USAF. (See Appendix 1, CEII-16.) The letter contained the following statement:

> While no credence is given to this report, previous experience with incidents of this nature dictates that USAF files should indicate that some official action had been initiated, in the event any official or public inquiries are made regarding this incident.

The incident itself concerns a reported sighting on April 14, 1957, at Vins sur Caramy, France. If taken out of context from other Close Encounter cases, it certainly does sound bizarre and must be given a high Strangeness Rating. One might even tend to excuse the attitude of the writer of the Air Force comment, Captain G. T. Gregory, then head of Blue Book. But it *cannot* be taken out of context. It is merely one of hundreds of reported Close Encounter cases. If one employs the shooting gallery technique of knocking off one duck at a time as it comes into range, one UFO case at a time without regard to its relation to similar cases, it is relatively easy to dismiss each as irrelevant and nonsensical. It is quite another thing if one becomes conscious not merely of one duck at a time but of the whole flock appearing in formation.

Here is a brief synopsis of the case itself. There reportedly landed on a road about 300 feet from two elderly French country women a curious metallic machine in the form of a big top about five feet tall. Just as it landed, a deafening rattle was heard coming from a metallic road sign some fifteen to twenty feet from the landing site. The sign had been set into violent vibration.

The cries of the women and the noise from the sign were heard by a man nearly 1,000 feet away. Thinking that there had been an accident, he went rushing down to them. He arrived in time to see the "top" jump off the road to a height of about twenty feet, turn, and land a second time, this time on another road, which forked from the first.

As it turned, it flew over a second road sign, and this one likewise vibrated violently, resonating as though it had been subjected to "violent shocks repeated at a rapid cadence." The machine, however, made no sound itself. It did not pass close to a third sign (presumably also metallic). This point assumes importance if one accepts the testimony of the local police, who, with the investigator, reportedly placed a compass near the two signs that had rattled and found a deviation of some fifteen degrees. Placed next to the Renault in which they had come, the compass showed a deviation of only four degrees, but there was no deviation at all near the sign that had not rattled.

Blue Book evaluated this case as Hoax without any evidence, presumably because they felt that was all that it could possibly have been. However, the French police adjutant in the area vouched for the integrity of the witnesses—"He affirmed at once that the witnesses are not only of good faith but they are above any suspicion of a hoax."

If one seeks a natural explanation for this case, I would suggest a purely meteorological explanation rather than a hoax. One might argue, although I don't, that a dust devil having the appearance of a top came by, rattled the signs, had a metallic appearance, selectively landed on a road each time, and "magnetized" the road signs.

Since I did not interview these observers myself, I can base judgment only on the French investigator's report (he seems to have done a good job of interrogation, as is evidenced by a reading of the full report). A dust devil simply does not fit the reported facts.

This is one of the few cases I have included in this book with which I did not have some measure of personal involvement. Perhaps the reader will wish to exclude it from the rest of the evidence, using it only as illustrative of official attitudes towards the UFO phenomenon. But the case fits in with the rest.

The prototype of the Close Encounter of the Second Kind stands out clearly from the selected cases just presented, but the reader and I must be disappointed that far more quantitative data are not available to portray it. Perhaps the Strangeness Rating of these cases is so great that they literally defy description that is translatable into the familiar quantitative terms in physics and astronomy. We meet an even more formidable category—as far as strangeness is concerned—in the next category, Close Encounters of the Third Kind.

CHAPTER TEN

Close Encounters of the Third Kind

I . . . waved. To our surprise the figure [on the UFO] did the same. . . . All missions boys made audible gasps. . . .

—FROM REVEREND GILL'S ACCOUNT OF THE SIGHTING AT BOAINAI, PAPUA, NEW GUINEA

We come now to the most bizarre and seemingly incredible aspect of the entire UFO phenomenon. To be frank, I would gladly omit this part if I could without offense to scientific integrity: Close Encounters of the Third Kind, those in which the presence of animated creatures is reported. (I say *animated* rather than *animate* to keep open the possibility of robots or something other than flesh and blood.) These creatures have been variously termed *occupants, humanoids, UFOnauts,* and even *UFOsapiens.*

Unfortunately one may not omit data simply because they may not be to one's liking or in line with one's preconceived notions. We balk at reports about occupants even though we might be willing to listen attentively to accounts of other UFO encounters. Why? In this *festival of absurdity,* as Aimé Michel has termed this part of the UFO phenomenon, why should a report of a car stopped on the highway by a blinding light from an unknown craft be any different in essential strangeness or absurdity from one of a craft from which two or three little animate creatures descend?

There is no logical reason, yet I confess to sharing a prejudice that is hard to explain. Is it the confrontation on the animate level that disturbs and repulses us? Perhaps as long as it is our own intelligence that contemplates the report of a machine, albeit strange, we still somehow feel superior in such contemplation. Encounters with animate beings, possibly with an intelligence of different order from ours, gives a new dimension to our atavistic fear of the unknown. It brings with it the specter of competition for territory, loss of planetary hegemony—fears that have deep roots.

Another thing bothers us: the humanoids seem to be able to breathe our air and to adapt to our air pressure and gravity with little difficulty. Something seems terribly wrong about that. This would imply that they must be from a place—another planet?—very much like our own. Perhaps our own? But how? Or are they robots, not needing to adapt to our environment?

Our common sense recoils at the very idea of humanoids and leads to much banter and ridicule and jokes about "little green men." They tend to throw the whole UFO concept into disrepute. Maybe UFOs could really exist, we say, but humanoids? And if these are truly figments of our imagination, then so must be the ordinary UFOs. But these are backed by so many reputable witnesses that we cannot accept them as simple misperceptions. Are then, *all* of these reporters of UFOs truly sick? If so, what is the sickness? Are these people all affected by some strange virus that does not attack sensible people? What a strange sickness this must be, attacking people in all walks of life, regardless of training or vocation, and making them, for a very limited period of time—only minutes sometimes—behave in a strange way and see things that are belied by the reliable and stable manner and actions they exhibit in the rest of their lives.

Or do humanoids and UFOs alike bespeak a parallel reality that for some reason manifests itself to some of us for very limited periods? But what would this reality be? Is there a philosopher in the house?

There are many such questions and much related information that is difficult to comprehend. The fact is, however, that the occupant encounters cannot be disregarded; they are too numerous. There is a surprising and highly provocative collection of literature on the subject of humanoids. In a catalog prepared by Jacques Vallee, which contains 1,247 Close Encounter

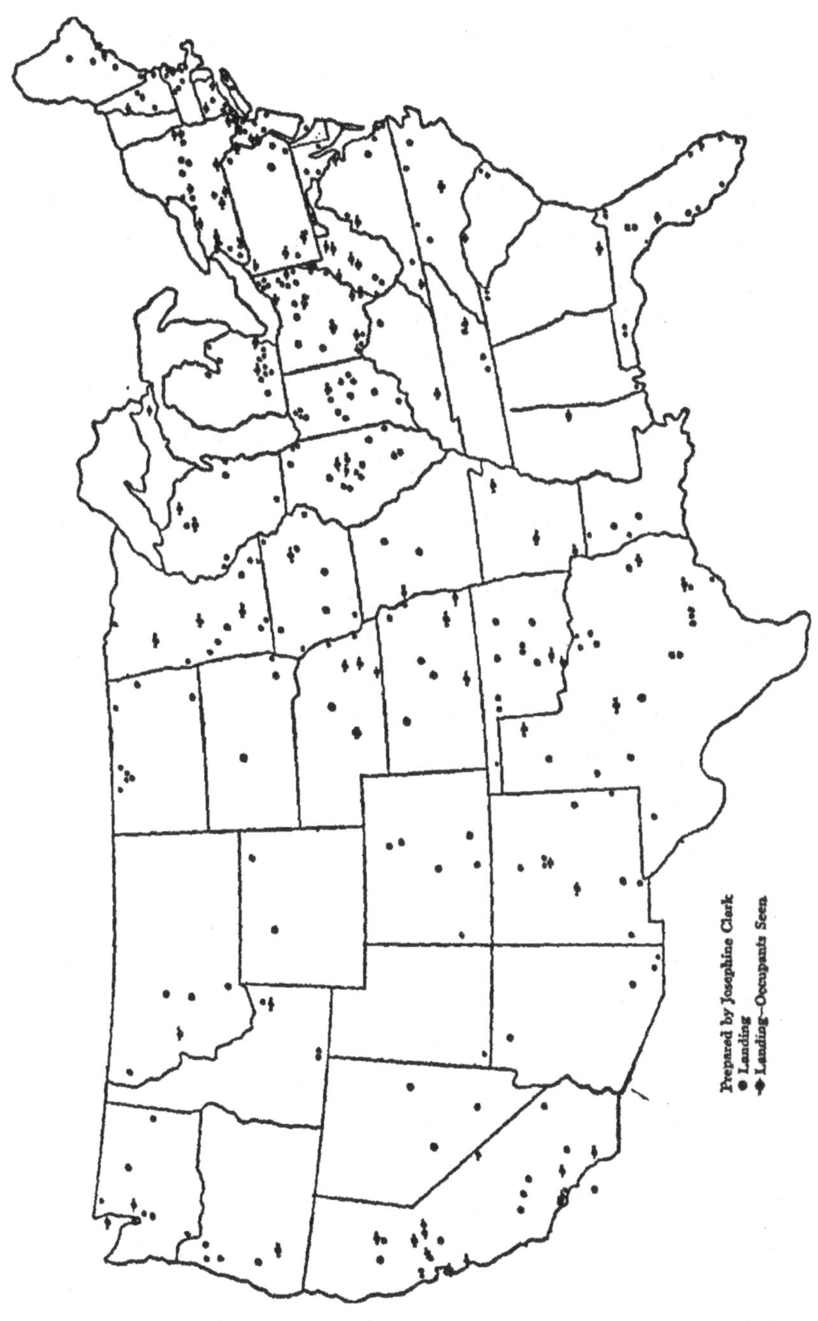
Prepared by Josephine Clark
Landing
Landing—Occupants Seen

cases, 750 are those in which an actual landing of a craft was reported. Of these, more than 300, or about 40 percent, were reported to have had humanoids seen in or about the landed craft. For both types, landings with or without humanoids, approximately one-third were multiple-witness cases.[1]

One would never suspect the magnitude of the problem of humanoids by perusal of the Air Force Blue Book files.[2] Out of the thousands of reports received by the Air Force, only forty-eight were reported UFO landings, and in only twelve of these did humanoids figure (for the years for which I have records available). During the same period 223 humanoid cases were reported from various areas of the world.

The manner in which the Air Force disposed of the few landing and still fewer humanoid cases is in itself of considerable interest.[3] Despite the widely held misconception that UFO reports of this sort are hallucinations, only two out of the forty-eight cases were attributed to hallucinations. Six were ascribed to the somewhat more vague category, Psychological; two, to Unreliable report, which in Blue Book terminology really means crazy; six were ascribed to hoaxes but on slim evidence. The majority was attributed *insufficient data*—a favorite term with Blue Book when it appeared that it would be too much trouble to acquire additional data.

Generally there was little or no follow-up in these cases. There were nine cases of reported landings of unknown aircraft, yet no attempt was made to ascertain further facts. Very real heights of creative evaluation were reached, however, in ascribing the famous Socorro, New Mexico, case to a ground light and the even more famous Hill case to radar inversion, a singularly inept evaluation inasmuch as no radar observations figured in that specific case at that time of night.

One may well ask why the Air Force received only twelve out of the sixty-five reported US cases of Close Encounters of the Third Kind or, for that matter, only forty-eight out of the 190 landing cases in the United States. I would surmise that many such cases "died" at the local air base to which they were reported. As we have seen, the standard Blue Book policy was to consider as "cases" only those incidents that came through official Air Force channels. Since landings and especially "little green men" were not to be taken seriously (by official policy), it is likely that the responsible officer at the local base deemed it unwise to transmit 'nonsense' messages.

It is hard to substantiate this surmise, yet from independent sources we know that UFO landings were reported to have occurred at Blaine Air Force Base (June 12, 1965), at Cannon Air Force Base, New Mexico (May 18, 1954), and at Deerwood Nike Base (September 29, 1957). None of these cases was transmitted to Blue Book, and if sightings made directly at air and military bases were not transmitted, it is quite likely that reports of landings—and especially of occupant cases—that were merely phoned in to the air base by a civilian were regarded as originating from a "nut" and hence not worth passing on through military channels. To Blue Book such a case would have been considered solved at the local level.

During my entire term as consultant to Blue Book I was asked to look into only two landing cases: the Socorro, New Mexico, case, which involved occupants (see Appendix 1, CEIII-1), and the Dexter, Michigan, case, which was a landing reported by only two of the many persons involved. I found the Socorro case extremely convincing and the Michigan case most unconvincing, even though the latter created greater public interest. I could, therefore, evade with ease this bizarre category entirely on the grounds that I choose to judge only those happenings with which I have had personal involvement, a rule I have followed, almost without exception, in the five other UFO categories. But despite my lack of personal acquaintance with Close Encounters of the Third Kind, I feel obligated to call attention to what has been reported in this category from around the world. I will do so by referring the reader to the many accounts available in the more scholarly UFO journals, already quoted, and to two books devoted entirely to this subject: *The Humanoids* (edited by Charles Bowen) and *Passport to Magonia* (by Jacques Vallee).

Readers will discover that there is a very great similarity in accounts of occupant cases in reports from over the world. They will learn that they are similar not only in the description of the appearance of most humanoids but in their reported actions. They will find the occupants reportedly picking up samples of earth and rocks and carrying them aboard their craft, much as US astronauts picked up moon rocks; they will find them seemingly exhibiting interest in human installations and vehicles; they will even find them making off with rabbits, dogs, and fertilizer!

It would be helpful, one feels, if we could demonstrate that Close Encounters of the Third Kind differ systematically from the other five UFO categories. Then we could, with some comfort, dismiss them. But they do not differ in any way—by geographical distribution, by times of occurrence, in numbers, and especially in kinds of observers—except that the relative number of cases with multiple witnesses is somewhat less (about one-third of the Third Kind cases have multiple witnesses) and that although the witnesses seem to represent the same cross section of the populace as those in the other categories, there are not as many observers having any degree of technical training. There are no pilots, air traffic control operators, radar operators, or scientists who have reported humanoids, according to my records. There are, however, people holding other types of responsible positions: clergy, police, electronics engineers, public servants, bank directors, military, miners, farmers, technicians, mail carriers, railroad engineers, medical doctors, and others gainfully and creditably employed.

Clearly, it is not only kooks who report humanoids. Indeed, I do not know of a report of this kind to have come from a person of demonstrated mental imbalance. Possibly there do not seem to be as *many* people of considerable technical training and sophistication who have made humanoid reports as have made other UFO reports because their very training and sophistication would naturally lead such people to be wary of exposing themselves to ridicule.

It appears, in short, that we cannot subdivide the UFO phenomenon, accepting some parts and rejecting others. We must study the entire phenomenon or none of it. Encounters of the Third Kind must in all fairness be included in this book.

Following as far as possible the policy I have adopted of discussing only those cases with which I have had personal involvement, I will choose those few cases with which I have had at least some peripheral involvement and for which I have been able to obtain some documentation. Unfortunately the most convincing case has come to me through private sources at the price of anonymity and hence cannot be fully discussed.

This case had four witnesses (see Appendix 1, CEIII-2), all family men holding responsible positions. At that time, two were engaged in work requiring military clearance, and their jobs would have been in severe jeopardy were their anonymity violated.* For the record, this reported event took place in North Dakota in November, 1961, in rain and sleet, late at night. The four men observed the landing of a lighted craft in a completely open and deserted field and, thinking that an aircraft was in serious trouble, stopped by the roadside, hopped the fence, and hurried toward what they judged to be the plane. Their surprise was understandably great when they discovered humanoids around the craft, one of which boldly waved them off in a threatening manner. One of the men fired a shot at the humanoid, which fell as if hurt. The craft soon took off, and the men fled.

The next day, although they reportedly had told no one of their bizarre experience, it was reported to me that one of the men was called out from work and led into the presence of men he had never seen before. They asked to be taken to his home, where they examined the clothing he had worn the night before, especially his boots, and left without any further word. To the best of my knowledge, none of the men involved heard further about the incident. There the mystery rests.

Two other cases, those of Temple, Oklahoma (March 23, 1966), and of Atlanta, Missouri (March 3, 1969), must be omitted by the rules of the game here; they are single-observer cases.

One case that should be excluded by these same rules is the extremely interesting event at Socorro, New Mexico, April 24, 1964 (see Appendix 1, CEIII-1), but since I have studied this case in some detail, I will touch on it.

Although there were other reported witnesses to the UFO, only Lonnie Zamora, a policeman in Socorro, was in a position to have seen the occupants. My original investigations, directed toward breaking apart Zamora's account by seeking mutual contradictions in it and also by seeking to establish

---

* I say this not to tantalize, but to emphasize to my scientific colleagues how difficult it is in this field to obtain data. In a sense we have regressed to the days before the founding of the Royal Society in England, when scientists had to sneak, so to speak, through back alleys as members of the "Invisible College."

Zamora as an unreliable witness, were fruitless. I was impressed by the high regard in which Zamora was held by his colleagues, and I personally am willing today to accept his testimony as genuine, particularly since it does fit a global pattern.

The Socorro incident is one of the classics of UFO literature, and a brief synopsis will suffice. On the afternoon of April 24, 1964, Zamora was on duty. At about 5:45 P.M., he was engaged in tracking a speeding motorist south of the town. He allowed the motorist to escape when his attention was drawn to a descending craft that was emitting a flame. At the same time he heard explosive sounds from the direction of the craft. While still some distance away, he was able to see the landed craft, which appeared like an up-ended automobile, and he noted the presence of two white-cloaked figures in its immediate vicinity.

He then lost sight of the object because of the hilly terrain and did not see it again until he rounded a curve and attained a higher elevation. To his astonishment he found himself looking down into a shallow gully at an egg-shaped metallic craft resting on legs extended from the craft. He was then less than 150 feet from the object, close enough to see a strange insignia on the side of the craft. Loud sounds from the interior of the craft caused Zamora to seek shelter as rapidly as he could. Glancing back over his shoulder at the craft, he saw it rise vertically and take off horizontally, disappearing shortly thereafter in the direction of Six Mile Canyon.

Zamora had already radioed his report into headquarters, and Sergeant Chavez was on his way. Had he not taken a wrong tum, the sergeant would have arrived in time to see the craft itself. As it was, he arrived to find a very shaken Zamora.

"What's the matter, Lonnie? You look like you've seen the devil," Chavez said.

"Maybe I have," replied Zamora.

I visited the site several days later and verified the landing marks and the charred plants. Chavez had, he told me in a long interview, verified the marks and the burned greasewood plants, which had still been smoldering at the time he first met Zamora at the site.

Measurements taken at the site showed that the diagonals of the quadrilateral formed by the four landing marks intersected almost exactly at right angles. One theorem in geometry states that if the diagonals of a quadrilateral intersect at right angles, the midpoints of the sides of the quadrilateral lie on the circumference of a circle, and it is thus of considerable interest that the center of the circle so formed virtually coincided with the principal burn mark on the ground. Under certain conditions the center of gravity of the craft would have been directly over the center of the circle, hence making the presence of the burn mark more significant.

On successive visits I continued my investigations into Zamora's credibility and traced the story of an unidentified witness who had stopped for gas just north of town. While at the gas station, he told of his encounter just south of town with a strange flying craft that was apparently in trouble and heading for a landing. He said that it must have been in trouble because he saw a squad car (Zamora's) going out across the sandy terrain toward it. He was unable to identify the craft as any normal aircraft.

I tried my best at the time to induce the Air Force to make an intelligence problem of finding the missing witness, but they evinced no interest whatsoever. At the time I thought that, had this been a federal case involving narcotics or counterfeiting, the FBI would certainly have located the missing witness. Because, it was merely a UFO case, the usual pattern of doing nothing was followed.

The Encounter of the Third Kind that had the largest number of witnesses was the sighting of June 26–27, 1959, centered about Reverend William Bruce Gill, an Anglican priest and a graduate of Brisbane University, who headed a mission in Boainai, Papua, New Guinea. (See Appendix 1, CEIII-3.) I first learned of the case in detail when I stopped at the British Air Ministry on an official visit from Blue Book in 1961.

I learned at that time that the British military view of the UFO problem was essentially the same as that of Blue Book; indeed, the British (and other governments as well) were looking to the US Air Force to solve the problem. I was told quite bluntly that with the funds and facilities available to the US Air Force there was little point to their doing anything about the problem, and they honestly felt that the US Air Force was doing something about it, but with negative results.

The British Air Ministry did not take Father Gill's sighting seriously, and almost with relief they gave me their report on it; it had apparently been cluttering up their files. Since then I have had access to a full report on this case[4] and have also been the recipient of a lengthy tape recording of a talk by Reverend Gill and, more recently, of an hour-long tape with Reverend Gill made by my colleague Fred Beckman.

Before judgment is passed on this affair, Reverend Gill should be heard. As a few excerpts from his tapes show, Reverend Gill is utterly sincere. He talks in a leisurely, scholarly way, delineating details slowly and carefully. The manner and content of the tapes are conducive to conviction. One would find it difficult to believe that an Anglican priest would concoct a story involving more than two dozen witnesses out of sheer intent to deceive. Critics of this case do not generally know that this report is only one of some sixty in the New Guinea area at approximately that time, all investigated by a colleague of Gill, the Reverend Norman Cruttwell, who has written a report covering the series,[5] only one of which, the case in point, involved humanoids.

The Department of Air, Commonwealth of Australia, however, was in doubt, although I have no record that they interviewed Father Gill in person. They wrote as follows to a colleague:

> CANBERRA ACT
>
> 28 January 1970
>
> Dear Sir,
>
> I refer to your letter dated 12 November 1969 concerning an unusual aerial sighting at Boainai, in Papua, New Guinea. The RAAF could come to no definite conclusion on the report, and inquiries with the United Kingdom and the United States could add no clues or answers.
>
> As a result these sightings have been classified as aerial phenomenon, but most probably they were reflections on a cloud of major lightsource of unknown origin.
>
> Yours Faithfully.

The letter is correct in one sense. When the brightly lighted UFO seen by Reverend Gill and his many colleagues went vertically upward, it illuminated the clouds as it passed through the overcast. The letter is quite correct, also, in stating that the light source was of unknown origin!

Here are a few excerpts from the report of the sightings in New Guinea. First from his notebook records made at the time of the sightings:

> The Boainai sightings climaxed a relatively short but remarkably acute period of UFO activity in the vicinity of east New Guinea. UFOs were observed by both Papuan natives and Europeans. Sightings were reported by educated Papuans and by totally illiterate natives relatively untouched by western civilization and quite ignorant of "flying saucers."

Next, from a letter written by Reverend Gill to a friend at a neighboring mission:

> Dear David,
>
> Have a look at this extraordinary data. I am almost convinced about the "visitation" theory. . . . I do not doubt the existence of these "things" (indeed I cannot now that I have seen one for myself), but my simple mind still requires scientific evidence before I can accept the from-outer-space theory. I am inclined to believe that probably many UFOs are more likely some form of electric phenomena or perhaps something brought about by the atom bomb explosions, etc. . . . It is all too difficult to understand for me; I prefer to wait for some bright boy to catch one to be exhibited in Martin Square. . . .
>
> Yours,
>
> Doubting William

The very next day this letter was written to the same person:

> Dear David,
>
> Life is strange, isn't it? Yesterday I wrote you a letter . . . expressing opinions re the UFOs. Now, less than twenty-four hours later, I have changed my views somewhat. Last night we at Boainai experienced about four hours of UFO activity, and there is no doubt whatsoever that they are handled by beings of some kind. At times it was

absolutely breathtaking. Here is the report. Please pass it around, but great care must be taken as I have no other. . . .

Cheers,

Convinced Bill

P.S. Do you think Port Moresby should know about this?

In a speech, Reverend Gill said:

> . . . and as I was about to turn around the corner of the house, something caught my eye in the sky, and I looked up toward the west. And there I saw at an angle of about forty-five degrees this huge light. I didn't think, of course, even then of flying saucers as such. I thought, well perhaps some people could imagine these things, but never me. And there it was. I called Eric Kodawara, and I said, "What do you see up there?" He said, "There seems to be a light." I said, "Well, you go and tell the teacher Steven Moi. Tell him to come along quickly." And then Eric went along, and he collected as many people as he could, and we all stood and gazed at it. Then we went up further up into the playing field, and the sighting went on. I've got it recorded here. I had decided by this time very quickly to get a notebook and pencil, and I thought, well, if anything is going to happen, it's going to happen now, and surely tomorrow I'll wake up and think it's been a dream, that I haven't really seen one. If I've got it down here in pencil, then I'll know at least I haven't been dreaming.

These are excerpts from the notebook recordings:

> Time 6:45 P.M. sky: patches of low clouds. Sighted bright white light, direction northwest. 6:50 called Steven and Eric. 6:52 Steven arrived confirms, not star. 6:55 send Eric to call people. One object on top moving—man. Now three men—moving, glowing, doing something on deck. Gone. 7:00 men 1 and 2 again. 7:04 gone again. 7:10 sky cloud ceiling covered sky height about 2,000 feet. Man 1, 3, 4, 2, (appeared in that order) thin electric blue spotlight. Men gone, spotlight still there. 7:12 men 1 and 2 appeared blue light. 7:20 spotlight off, men go. 7:20 UFO goes through cloud. 8:28 clear sky here, heavy cloud over Dogura. UFO seen by me overhead. Called station people. Appeared to descend, get bigger. 8:29 second UFO seen over sea—hovering at times. 8:35 another over Wadobuna Village. 8:50 clouds

> forming again. Big one stationary and larger. Others coming and going through clouds. As they descend through cloud, light reflected like large halo on to cloud—no more than 2,000 feet, probably less. All UFOs very clear. "Mother" ship still large, clear, stationary. 9:05 clouds patchy, numbers 2, 3, 4 gone. 9:10 number 1 gone overhead into cloud. 9:20 "Mother" back. 9:30 "Mother" gone, gone across sea toward Giwa. 9:46 overhead UFO reappears, is hovering. 10:00 still stationary. 10:10 hovering, gone behind cloud. 10:30 very high hovering in clear patch of sky between clouds. 10:50 very overcast, no sign of UFO. 11:40 heavy rain. Data sheet of observation of UFOs 6:45-11:04 P.M. Signed William B. Gill.

Reverend Gill's narrative account contains this information:

> 7:12, men 1 and 2 appeared—blue light. I might mention here. that the cloud ceiling was about 2,000 feet, and I judged the cloud ceiling by a mountain. And all of this, of course, was well under the cloud ceiling. By this time, in a space of twenty-five minutes, the sky had clouded over. At 7:20 the UFO went though the clouds, right through. At 8:28 the sky was beginning to clear again, although it was heavy, the cloud cover was heavy over Giwa. UFO seen by me now over it. I called the station people the second time that night around twenty-eight minutes past eight, and it appeared to descend and get bigger. . . . Others were coming and going through the clouds—remember we now had patches of clouds. They were descending through the clouds and the glow of the discs was reflected at the base of the clouds, and then they would go in through the cloud again, and they seemed to enjoy doing that.
>
> Then came the next night, and this was the interesting one. A large UFO was first sighted by one of the nurses at the hospital at 6:00 P.M. . . . It happened this way: we were walking, and this thing came down to what we estimated as the closest we had seen it, and I was practically the closest we were ever to see it. Somewhere between 300 and 500 feet it dropped down. It was not dark, and we could see it quite clearly. It was still bright and sparkling, but it seemed very near and clear. And there was this figure again on the decking, as I called it, at the top. And it was the teacher who said, "I wonder if it is going to land on the playing field." I said, "Why not?" And so we

> waved, like that,—Hello—and we were a bit surprised now, and the thing waved back. And then Eric, who was with me, my constant companion, waved his two arms, along with another lad, and then the figures waved two arms back.

Although I did not personally investigate the Gill New Guinea case, now one of the classics, I am impressed by the quality and number of the witnesses and by the character and demeanor of Reverend Gill as revealed by his report and tapes.

The self-styled "arch enemy of UFOs," Dr. Donald Menzel, of Harvard, has taken a characteristic opposite view. In his *Analysis of the Papua-Father Gill Case* (see Appendix 2), he dismisses the entire case as a sighting of Venus under the hypothesis that Reverend Gill was not wearing his glasses at the time. Unfortunately, he neglected to ascertain the following: the UFO at times was seen under cloud cover; Venus was pointed out separately by Gill; and Reverend Gill was wearing properly corrected glasses at the time.

Another classic Close Encounter of the Third Kind is the Kelly-Hopkinsville sighting of August 21, 1955 (see Appendix 1, CEIII-4), in which it was the humanoids who took the center stage, the UFO being mentioned only in passing. My connection with this affair was purely fortuitous since I had not been called in to consult on this case. A few months after it had occurred I was engaged in setting up twelve satellite tracking stations around the world, with very little time for UFO investigations. It so happened, however, that one of the electronics technicians I had hired to work on the crystal-clock timing mechanisms of the Baker-Nunn satellite cameras was one Bud Ledwith, who, I learned later, had been an engineer and announcer at Radio Station WHOP in Hopkinsville, Kentucky. On the morning immediately after the Kelly event, Ledwith began a detailed investigation of his own. From him I obtained the full story which included signed affidavits and sketches.

The Kelly-Hopkinsville case, if considered entirely apart from the total pattern of UFO sightings, seems clearly preposterous, even to offend common sense. The latter, however, has not proved a sure guide in the past history of science. Blue Book records on this event are sketchy, and little or no investigation was conducted. Still, the case is carried in Blue Book files as Unidentified. That much it certainly is.

Ledwith interrogated all seven adult witnesses and asked each to draw his recollection of what the occupants looked like. Signed statements were obtained from each adult witness. Ledwith then made a composite drawing of the occupants and had the witnesses sign that. He kindly turned over to me his files, including the original drawings and notes, and has given me permission to make full use of them.

Since this classic case has been treated elsewhere, a brief synopsis will suffice here: A conventional UFO was seen by only one witness to land in a gully near the farmhouse occupied by the Sutton family. This witness, coming back to the farmhouse, told of his sighting. His report was promptly discounted, and he was subjected to mild ridicule. Less than an hour later the occupants of the house were alerted by the violent barking of the dog in the yard. Two of the men in the house went to the back door to see who was coming. A small "glowing" man with extremely large eyes, his arms extended over his head "as though he were being robbed," slowly approached the house.

In that area some of the country people in the economic and social framework of these witnesses "shoot first and ask questions later." This is precisely what the two Sutton men reportedly did, one with a .22 rifle and the other with a shotgun. Both men fired when the "UFOnaut" was about twenty feet away from the house; the sound was described "just like I'd shot into a bucket." The visitor did a quick flip and scurried away into the darkness.

Soon another visitor appeared at the window and was promptly fired at through the window. The screen bears the souvenir (a bullet hole) of this attack on the invader. Going out to see if they had killed the intruder, those behind the first man saw, as he momentarily stopped under a small overhang of the roof, a claw-like hand reach down and touch his hair. Gunplay again, directed toward the creature on the roof and toward one suddenly noted in a nearby tree branch. The latter was apparently hit directly, but it *floated* to the ground (a maneuver that was to be repeated) and scurried away.

Unnerved by the ineffectiveness of the guns, the entire family was soon confined within the house behind bolted doors. From time to time the visitors reappeared at the windows.

After about three hours the family decided that they had had enough of this one-sided siege, and, apparently making a quick decision, all eleven of the occupants of the house piled into two cars and headed into town to the police. That was at eleven in the evening. Since it takes serious provocation to pack eleven people into cars late at night and to rush the seven or so miles to the police station, asking for help, it gives some indication of the terror that must have gripped the family by this time.

When the family returned, the police surveyed the territory, lights flashing amidst considerable commotion, but nothing was found.

After the police left and all was once again dark and quiet, the creatures reappeared, according to all witnesses.

Ledwith's account[6] of how the description of the little creatures was obtained is germane:

> When I got there [to the radio station] everyone greeted me with, "Have you seen the little green men yet?" I inquired and got a vague account of the night's happenings. Remembering a magazine article that I had read not long before, about the way the police artists reconstruct facial features from witnesses' descriptions, I decided to clarify the situation. I had once studied art and thought I might be able to get sketches.
>
> With me I took one of the men at the station, in order to have a witness who could watch and listen and make sure that I was not leading or guiding the people I interviewed.
>
> Despite the events of the previous night the men of the house had gone out of town on their planned business for the day and were not in. The womenfolk agreed to an interview.
>
> I did not lead the women in any way as the pictures were drawn. They were extremely positive of what they had seen and had not seen; it was a matter of following the directions as to the shape of the face, eyes, hands, and body. If I even so much as advanced a supposition of how one particular feature might have looked, they would quickly

> correct me. . . . It wasn't long before the "apparition" began to take form. The eyes were like saucers, large and set about six inches apart; they seemed to be halfway around the side of the face. . . . The head itself was circular and completely bald on top. . . . We progressed to the body. No one was sure whether there was a neck or not, so we left it out. According to the women, the body was thin, with a formless straight figure. . . . The arms were peculiar; they were almost twice as long as the legs . . . the hands were huge, bulky looking things. . . . The only part of the face that no one could describe was the nose. . . . I tried to sketch in a nose . . . but no one was sure, so we removed it.

After spending three hours interviewing the women of the household, Ledwith and his companion received permission to return that evening, when the men would be home. They did so, arriving shortly before the men did. When the men came home, "Cars were lined up for half a mile in both directions." When "Lucky" Sutton, conceded to be the dominant personality in the household, arrived on the scene, "He came into the house like a bear."

> His eyes dropped to the table, where I had placed my drawing. Without saying another word he sat down, . . . looked it over . . . and said, "No, the face is almost round, it doesn't come to a point." We got right to work on the men's drawing, using the women's as a guide and making changes as the three men indicated. . . . The mouth was disputable; Lucky was adamant that there had not been any mouth at all. If any, it was not much more than a straight line across the face. To pacify those who had seen it, I drew in a straight line, high, from ear to ear.
>
> Those seven people had given me almost parallel stories and almost identical pictures. It would be impossible for so many people to give me false accounts and pictures that tallied so closely unless they first talked together and decided what each feature [and event] looked like; but three of the men had left very early that morning for Evansville and had not been home throughout the day. These were not interviews in which one person would look at another and say, "Is that what you thought it looked like?" No, all seven were sure of what they had seen, and no one would retract a

> statement . . . even under close cross-examination. I use that word loosely where Lucky is concerned; you don't exactly cross-examine Lucky Sutton.
>
> . . . as the report spread outside the family, they were distorted in all directions; everyone who told the story seemed to add his own ideas of how the creatures looked. For this reason I am pleased that we had the advantage of time. Our morning interview was the first complete report of the whole night's happenings. The women were friendly and relaxed and we had no disturbance. The sight-seeing horde had not yet become overwhelming. That night we talked to the men in the same way, immediately after they came home, before they had had any opportunity to discuss the first interview with the others. I was greatly impressed with their sincerity, both the men and the women. . . .

The participants in this case received so much adverse publicity and personal harassment that they soon refused to discuss the matter with anyone, making further meetings difficult. However, one successful follow-up was made nearly a year later by one of the most sincere and dedicated UFO investigators I have met, Isabel L. Davis, of New York City, who privately made a trip to Kelly. Under the influence of her quiet yet determined personality many of the original witnesses were persuaded to review and discuss in great detail the events of August 21–22, 1955.

Isabel Davis has written a full account of her visit, an excellent document worthy of publication, and has kindly furnished me with a copy of her manuscript, which, in the main, fully supports the earlier investigations of Ledwith. Considered together, these accounts give us a picture of a truly bizarre and, in ordinary terms, completely unexplainable event.

Seven adults and four children attested to the essentials of the event. The witnesses were not "status inconsistent," and that theory, propounded by D. I. Warren,[7] does not account for this sighting nor for a great many others. Warren maintains that UFO reports are more apt to come from people whose economic status is not consistent with their intellectual capacity and training: for example, a poorly trained person occupying a relatively high economic and social status, or vice-versa.

I would not have given the Kelly-Hopkinsville case this much attention were it not for the fact that I know the principal investigators, Ledwith and Davis, well, particularly Ledwith since he was in my direct employ for nearly two years on the satellite tracking program.

There is an even greater reason: the humanoids are themselves a prototype that has occurred again and again throughout the years, going back, as Vallee so convincingly points out in *Passport to Magonia*, to the myths and legends of many cultures. It is highly improbable that the Suttons, "who did not have telephone, radio, television, books, or much furniture," were aware of UFO lore and could have known that many times in the past creatures like those they had delineated had been described. The resemblance to the "little people" described by many cultures is striking.

We are not, of course, justified in concluding that the Kelly creatures stemmed from the imagination alone or, conversely, that the source of ancient legends lies in the actual appearances of such creatures in the past or that real humanoids were seen. As in other aspects of the entire UFO phenomenon, the call is clearly for more study.

The Suttons themselves were convinced that they had had a real experience, a pattern of reaction I have found consistently. Let the report of Isabel Davis underscore this:

> Finally, the Suttons stuck to their story. Stubbornly, angrily, they insisted they were telling the truth. Neither adults nor children so much as hinted at the possibility of a lie or mistake—in public or to relatives; there was no trace of retraction.

Davis further remarks on the absence of "protective rationalization" used by UFO sighters, who, though personally convinced, wish to remain in the good graces of their fellows by saying something such as, "Of course, it must have been an airplane. . . . I could have been mistaken"—accompanying their disclaimers by an embarrassed laugh or giggle. As she states:

> The Suttons seem never to have been tempted to recant and get back into the good graces of society. . . . Their costly refusal to give an inch to skepticism may not prove anything about the truth of their story, but it does tell us something about them.

It may be interjected that the Kelly case had only one witness as far as the UFO itself was concerned, even though eleven people witnessed the occupants. It was thus a close encounter with occupants rather than with a craft. In many cases in UFO records the occupants have been reportedly clearly sighted but their craft viewed only for a moment. In the Socorro, New Mexico, case, of course, just the opposite was true. Zamora saw the occupants only from a distance, but the craft close at hand.

I now introduce one "contactee case," not because I accept the usual contactee reporter but because it is not a contactee case in the usual sense: it has no pseudoreligious/UFO cult overtones, no platitudinous cosmic messages of little content. Moreover, it was a most thoroughly investigated case study and the subject of the book *The Interrupted Journey* by John Fuller. It is the story of Betty and Barney Hill.

I do not, however, feel that it fits the pattern of the Close Encounters of the Third Kind, so is not useful in contributing to the prototype of this category. I include it, in a sense, to demonstrate the contrast between it and the usual contactee story and the typical Close Encounter of the Third Kind. In addition, the case is well known to the public, and it created great interest.*

The Betty and Barney Hill case—of the racially mixed New Hampshire couple who, under repeated independent hypnotic treatment by Boston psychiatrist Dr. Benjamin Simon, related a story of an Encounter of the Third Kind, in which they were abducted aboard a "spaceship"—is one that naturally created tremendous interest; while the story is fully covered in Fuller's book, a brief synopsis is needed here. (See Appendix 1, CEIII-5.)

Returning along a lonely road late at night from a Canadian vacation, Betty and Barney Hill espied a descending UFO. Eventually it landed and constituted a roadblock to the progress of their journey. The couple was approached by humanoids.

Some two hours later the Hills found themselves thirty-five miles farther along on their journey but with no recollection of what had happened

* As a result of Fuller's book and its condensation in a national magazine, it is one of the most publicized cases. In popular discussions I rarely fail to receive questions about it.

during those two hours. This amnesia continued to bother them, leading to physical and mental disorders, and they finally were referred to Dr. Simon, whose success with amnesia cases is well known.

Under repeated hypnosis they *independently* revealed what had supposedly happened. The two stories agreed in considerable detail, although neither Betty nor Barney was privy to what the other had said under hypnosis until much later.

Under hypnosis they stated that they had been taken separately aboard the craft, treated well by the occupants, rather as humans might treat experimental animals, and then released after having been given the hypnotic suggestion that they would remember nothing of that particular experience. The method of their release supposedly accounted for the amnesia, which was apparently broken only by counter-hypnosis.

The medical experiments involved inserting a needle into Betty Hill's navel and collecting nail parings and skin shavings. In one "amusing" incident Barney's false teeth were removed; the attempt to do the same with Betty, who has her own teeth, of course failed miserably.[†]

Shortly after the publication of Fuller's book he and I were invited to dine with the Hills at the home of Dr. Simon, outside Boston. By previous agreement of all parties, Dr. Simon put the Hills into a hypnotic trance and allowed me to question them while they were under hypnosis. This proved to be quite an experience for me, for as Barney described the abduction aboard the craft he became emotionally disturbed, and Dr. Simon had difficulty in keeping him calm.

The emotional content of the Hills' experience came through loud and clear, but all things considered, the information content of the one and one-half hour session was minimal. Part of this inadequacy was undoubtedly due to my inexperience in questioning anyone under hypnosis. At times both Betty and Barney spoke haltingly except at emotionally charged moments. Direct questions were often answered gropingly, reminding me at times

[†] One can imagine a learned paper presented at a scientific meeting on "planet X" in which it is described that their expedition to Earth disclosed that male black people have teeth that can be removed but white females have teeth which do not come out!

of the difficulty of obtaining information from persons who are seriously ill; there were long pauses. However, at no point did I gain the impression that there was any deliberate attempt to avoid giving information. Later at dinner, the Hills were spritely, charming, and talkative. There was no question of their normalcy and sanity.

A few excerpts from my hypnotic session with them will illustrate both the intensity of the emotional experience revealed by hypnosis and the very apparent sincerity of the subjects.

DR. SIMON: All right, now we're coming back in time to the end of that trip to Niagara Falls, when you're coming back and had the experience with the unidentified flying object. What is your feeling now? Were you abducted or weren't you?

BARNEY: I feel I was abducted.

DR. SIMON: Were you abducted?

BARNEY: Yes. I don't want to believe I was abducted, so I say I feel because this makes it comfortable for me to accept something I don't want to accept that happened.

DR. SIMON: What would make it comfortable?

BARNEY: For me to say I feel.

DR. SIMON: I see. Why are you uncomfortable about it?

BARNEY: Because it is such a weird story. If anyone else told me that this had happened to them, I would not believe them, and I hate very badly to be accused of something that I didn't do when I know I didn't do it.

DR. SIMON: Now what is it you are accused of?

BARNEY: If I am not believed that I have done something and I know I have done it.

DR. SIMON: Well, suppose you had just absorbed Betty's dream.

BARNEY: I would like that.

DR. SIMON: You would like that; could that be true?

BARNEY: No. . . .

BARNEY: (shouting): I didn't like them putting that on me! I didn't like them touching me!

DR. SIMON: All right, all right. They're not touching you now, they're not touching you at all. We'll let that go.

Now Dr. Hynek is going to talk to you, and Mr. Fuller may talk to you, and you will both carry out their instructions as if they were mine for this time. You'll answer all the questions that may be put to you and carry out any instructions given by the three of us while you're in this trance. But after this you will respond only to me.

HYNEK: Barney, you will remember everything clearly, and I want you to tell me what is happening; you have just heard the beep-beep-beep; I want you to tell me what it sounded like, and then each of you just relive and tell me what is happening as you are driving down the road.

Suddenly Barney takes up the narrative, presumably at the point at which they encountered the humanoids:

BARNEY: Betty, it's out there—it's out there, Betty! Oh God, this is crazy. I'm going across a bridge—I'm not on Route 3. Oh, my! Oh, my! Oh, my! [Barney breathing very heavily]. Oh, I don't believe it. There are men in the road. I don't believe it. I don't want to go on. It can't be there. It's the moon.

DR. SIMON: Go on, Barney. You remember everything clearly—everything's clear.

BARNEY: I'm out of the car, and I'm going down the road into the woods. There's an orange glow; there's something there. Oh, oh, if only I had my gun; if only I had my gun [in an excited, despairing tone]. We go up the ramp. I'd love to lash out, but I can't. I'd love to strike out, but I can't. My emotions—I got to strike out—I got to strike out! . . . My feet just bumped, and I'm in a corridor. I don't want to go. I don't know where Betty is. I'm not harmed; I won't strike out, but I will strike out if I'm harmed in any way. I'm numb. I have no feeling in my fingers. My legs are numb. I'm on the table!

DR. SIMON: It's all right. You can stop there. You're on the table, but you're quiet and relaxed and you just rest now until I say, "Listen, Barney." You won't hear anything I'm saying for a little while. Betty, what's going on?

BETTY: We're riding—Barney puts on the brakes, and they squeal, and he turns to the left very sharply, I don't know why he's doing this. We're going to be lost in the woods. We go around a curve [Pause.] Barney keeps trying to start it—it won't start. In the woods now they come up to us. There's something about the first man who's coming up. This is when I get frightened, and I gotta get out of the car and run and hide in the woods. . . .

DR. SIMON: Stop, Betty, stop for a moment. You don't want to hear anything I say.

There is an interlude here in which Barney cries out in a very distraught manner, and Dr. Simon works very hard to calm him down. Finally he turns to Betty again.

DR. SIMON: Betty, you can hear me now.

BETTY: Yes.

DR. SIMON: Go on.

BETTY: I want to open the car door now and get out and run and hide in the woods. . . .

HYNEK: Have you ever seen anything before that even resembled this?

BETTY: No.

HYNEK: Was the moon shining down on it? Could you see the moon at the same time?

BETTY: It was a very moonlit night. It wasn't quite as clear as daylight, but I could see. It was on the ground, and there was like a rim around the edge.

HYNEK: Was it resting on legs or was it flat on the ground?

BETTY: The rim was a little bit above the ground, and there was a ramp that came down.

HYNEK: How big was it, Betty? Compare it to something you know, Betty, in size.

BETTY: I tried to think it. . . .

HYNEK: How about a railroad car? Was it bigger than a railroad car or smaller than a railroad car?

BETTY: I can't picture the size of a railroad car. I would say if it landed out here on the street—let's see, it would go from the corner by the front of the house and it would go beyond the garage.

HYNEK: What were your thoughts as you came closer and closer to it?

BETTY: To get the h out of there if I could.

HYNEK: And why couldn't you?

BETTY: I couldn't seem to. I—their man was beside me. All I could say was, "Barney, Barney, wake up." He asked me if his name was Barney. I didn't answer him 'cause I didn't think it was any of his business. And then when we got—I saw this—I knew they were gonna want us to go on it. I didn't want to go. I kept telling them I'm not gonna go—I don't want to go. And he said for me to go ahead, go, that they just wanted to do some simple tests. As soon as they were over with, I'd go back to the car.

HYNEK: Did they tell you where they were from?

BETTY: No.

HYNEK: What kind of sounds did they make?

BETTY: They were like—words—like sounds of words.

HYNEK: English words?

BETTY: No.

HYNEK: But you understood them?

BETTY: Yes.

HYNEK: How do you explain that?

BETTY: It was—all I can think of is—learning French.

HYNEK: Learning French?

BETTY: Yes.

HYNEK: Do you think it was French?

BETTY: No, but it was like learning French. When you first hear the French word, you think of it in English.

HYNEK: I see. So you heard these sounds in some language, and you understood them as if they were English, is that it?

Dr. Simon touched Betty's head. "I touch your head now, and you'll be resting and relaxed, and you'll not hear anything further until I touch your head again," he said. "You won't hear anything further. Barney, you can hear me now, you're comfortable and relaxed. You have told me that you have gone into the vehicle, is that right?"

BARNEY: Yes.

DR. SIMON: They had taken you in, and they, had put you on a table.

BARNEY: Yes.

DR. SIMON: And they talked to you, is that right?

BARNEY: Yes.

DR. SIMON: Tell us how they talked; answer Dr. Hynek on that.

HYNEK: Did you see them, Barney, open their mouths and, and, if so, how wide did they open their mouths?

BARNEY: Their mouths moved, and I could see them.

HYNEK: Try to tell me what the sounds were or if they represent anything you know. Is there any animal that you can think about that makes a sound like what they were making,

BARNEY: No.

HYNEK: What were the sounds like? [Quivering *oh! oh! oh!* sounds come from Barney]

HYNEK: What did you think about them, or did you think about them at all?

BARNEY: I thought if only I could haul my fists up. . . .

HYNEK: This is while you were on the table?

BARNEY: Yes. I wanted to fight. I didn't know where Betty was, and yet every time I would move or struggle this bright light in my head would keep me calm.

The fact that the major portion of the Hills case was revealed only under hypnosis makes it atypical. The first part of the Hills' case starts out just like many others—a light in the sky that grows larger and brighter as it approaches, the landing, and the frightening experience of the encounter. But the abduction, the physical tests, and the entire sequence of events until they found themselves thirty-five miles down the road two hours later, with complete amnesia during those two hours, is atypical.

If we discount entirely the account revealed only under hypnosis, the first portion fits the pattern. The atypical portion is not amenable to study except as an atypical event. When and if other cases of hypnotic revelation of close encounters become available for study (one recalls that the Hills waited several years before seeking treatment), we will be able to note whether they also form a pattern.

What of the occupants themselves? They seem to come in two sizes; large and small, with the former predominating. The Hopkinsville humanoids and many of those recounted in *Passport to Magonia* by Jacques Vallee are much akin in appearance to the "little folk" of legend and story—elves, brownies, etc. Large heads, spindly feet, and, generally, a head that sits squat on the shoulders without much evidence of neck are often described. The larger humanoids are reported to be human size or a little larger and are generally very well formed. Sometimes they have been termed beautiful. The smaller ones generally are described as about three and a half feet tall.

But this is not the place to attempt a taxonomy of humanoids; the reader will do well to refer to Bowen's *Humanoids* and to Vallee's *Passport to Magonia.*

What, finally, can be said of Close Encounters of the Third Kind? They differ from other close encounter cases only by definition, by the reported presence of occupants, and by the fact that these encounters are not as frequently reported by highly trained and sophisticated people as are other close encounters. Whether these experiences occur to such people is, of course, open; unless they report such incidents they virtually do not exist and certainly cannot be studied.

Circumstances dictated that in my work over the past twenty years I did not come into direct contact with many cases in this category; largely for that reason I prefer, in my own thinking, to rest the UFO problem on the prototypes of the first 5 categories of UFOs: Nocturnal Lights, Daylight Discs, Radar-Visual, and Close Encounters of the First and Second Kinds. I feel that I have a greater grasp on these groups because of my work with the various reporters of cases in those five categories. Therefore I must leave it to the reader's own judgment what weight to assign to Close Encounters of the Third Kind in assessing the whole problem, always remembering that it may yet be discovered that the humanoid cases are the key to the whole problem.

PART III

Where Do We Go from Here?

INTRODUCTION: THE INVISIBLE COLLEGE

Now that you, the reader, have an overview, albeit introductory, of the UFO phenomenon and particularly of the data of the problem, you can come to appreciate and, I expect, to deplore the manner in which the problem has been handled over the past score of years or more. In this section we shall first survey the manner in which the Air Force publicly handled the problem, next the manner in which the Condon committee at the University of Colorado treated it. Finally, I shall suggest a positive approach to the problem.

As one becomes familiar with the wealth of material in this field and the manner in which it has been handled, he will feel, as I have, the frustration a person might experience in describing the colors of a sunset to one born blind. The blind person has not deliberately assumed his blindness but it would seem that the world of science has placed blinders on the eyes of science, but not entirely without some good reason. The confusion surrounding the subject has been great, and its "sponsorship" all too frequently by persons ill equipped to assess and treat it critically. These factors in addition to the "lunatic fringe" were sufficient to cause most scientists to avoid the subject of UFOs.

A major first step in removing the scientific blinders has, in my opinion, been the statement of the Special Committee of the American Institute of Aeronautics and Astronautics (December, 1968) and the publication of UFO case studies in their *Journal of Astronautics and Aeronautics.* These clearly indicate a challenge to human curiosity and thus to the scientist, although to what scientific discipline it applies is not clear. At the moment the problem belongs exclusively to the physicist, the sociologist, the psychologist, and even to the student of the occult.

I have positive evidence from personal correspondence and conversations with scientists that their interest is increasing but that it is still, in most cases, anonymous. There is truly a growing Invisible College of scientifically and technically trained persons who are intrigued by the UFO phenomenon and who, if provided with opportunity, time, and facilities, are most willing to undertake its serious study. They represent an international group ready to accept the challenge of the UFO.

CHAPTER ELEVEN

The Air Force and the UFO—Pages from Blue Book

DEPARTMENT OF THE AIR FORCE
HEADQUARTERS FOREIGN TECHNOLOGY DIVISION (AFSC)
WRIGHT-PATTERSON AIR FORCE BASE, OHIO 45433

OFFICE OF THE COMMANDER

6 SEP 1968

Dr J Allen Hynek
Dearborn Observatory
Northwestern University
Evanston, Illinois 60201

1. During the past few years you have publicly criticized Project Blue Book for their lack of scientific evaluations of some unidentified flying object reports.

2. I would like for you to address your efforts, during the next thirty days, towards defining those areas of scientific weakness which presently exist in the Project Blue Book office. Please confine your paper to the scientific methodology which should be used and do not concern yourself with Air Force policy or history.

3. Your recommendations should be precise, detailed, and practical. Your report should reach my office no later than October 1, 1968.

RAYMOND S. SLEEPER, Colonel, USAF
Commander

The above letter marked the first time in my twenty years of association with the Air Force as scientific consultant that I had been officially asked for criticism and advice on scientific methodology and its application to the UFO problem. It is quite true that for some time before the receipt of the above letter I had become increasingly critical of Blue Book, but it had apparently escaped official attention and action. Now it was clear that, at long last, attention was being paid.

I was, of course, pleased to have an official audience, and I answered in full (see Appendix 5). To place my reply in proper perspective it is necessary to trace briefly the history of Blue Book and my association with it and to see clearly the part both the Pentagon and the scientific fraternity played in shaping and crystallizing Blue Book posture toward the UFO problem. It will be noted that Colonel Sleeper's letter did not include a request for comments on policy. This continued to be determined, as it had in the past, at much higher levels, through Pentagon channels.

Fifteen years earlier the prestigious Robertson panel[1] had labored for parts of five days (January 14 to 18, 1953) and had brought forth these conclusions and policy recommendations:

> (a) That the evidence presented on Unidentified Flying Objects shows no indication that these phenomena constitute a direct physical threat to national security. We firmly believe that there is no residuum of cases which indicates phenomena that are attributable to foreign artifacts capable of hostile acts, and that there is no evidence that the phenomena indicate a need for the revision of current scientific concepts.
>
> (b) That the continued emphasis on the reporting of these phenomena does, in these parlous times, result in a threat to the orderly functioning of the protective organs of the body politic. We cite as examples the clogging of channels of communication by irrelevant reports, the danger of being led by continued false alarms to ignore real indications of hostile action, and the cultivation of a morbid national psychology in which skillful hostile propaganda could induce hysterical behavior and harmful distrust of duly constituted authority.

The panel recommended:

> That the national security agencies take immediate steps to strip the Unidentified Flying Objects of the special status they have been given and the aura of mystery they have unfortunately acquired.

> That the national security agencies institute policies on intelligence, training, and public education designed to prepare the material defenses and the morale of the country to recognize most promptly and to react most effectively to true indications of hostile intent or action.

It would seem that the panel's attention was directed largely to a defense and security problem rather than to a scientific one. This could have been expected in a sense since the meeting had been called by and they had been instructed by the CIA. No mention was made of or explanations offered for the great many Unidentified cases already in the Blue Book files. Since the cases had been selected for them by Blue Book, which already had stated views on the subject of UFOs, the prejudicial nature of the "trial of the UFOs" is obvious. The august panel members were examples of the old saying, "When you can keep your head when all about are losing theirs, you don't understand the situation." The panel was not given access to many of the truly puzzling cases.

At the time the panel was called into existence, the Battelle Memorial Institute, of Columbus, Ohio, was engaged in a statistical study (which eventually appeared as Blue Book Report No. 14, a remarkable document if one reads between the lines), and in a proper scientific spirit the officers of Battelle had pointed out* that there was a distant lack of reliable data and that even the well-documented reports presented an element of doubt about the data. They called for an upgrading of the data before any broad policy decisions were made, and they implied (though they were too diplomatic to say so) that the whole Robertson panel was premature and not likely to get anyplace. The Robertson panel did get someplace: they made the subject of UFOs scientifically unrespectable, and for nearly twenty years not enough attention was paid to the subject to acquire the kind of data needed even to decide the nature of the UFO phenomenon. Air force public relations in this area was egregious, and the public was left with its own decisions to make: was the Air Force attitude a result of cover-up or of foul-up and confusion?

---

* A letter of January 9, 1953, addressed to the attention of Captain E. J. Ruppelt, first head of Blue Book.

The Air Force officially entered the flying saucer arena on January 22, 1948, in response to an exchange of letters, in the latter part of 1947, between the commanding general of what were then the army Air Forces and the chief of the Air Material Command, Air Technical Intelligence Center (ATIC) at Wright-Patterson Air Force Base, in Dayton, Ohio.

The events of the summer of 1947 had been disturbing: too many reports of strange objects seen in the sky had been made by seemingly responsible people—mountain pilots, policemen, commercial pilots, military observers, etc. Charged with the defense of the country from the air, the Air Force had become instantly concerned. The first explanation to come to mind was, of course, that a foreign power had developed a new and potentially sinister device posing an obvious threat to our security. As frightening as this might seem, it was nonetheless a concept that the military mind could immediately grasp and with which it felt it could come to grips. Foreign technology intelligence investigations were right up the Air Force's alley. And thus Project Sign, sometimes incorrectly referred to as "Project Saucer," was born. Its staff went to work to examine critically the first series of reports, and very shortly thereafter I was asked to check on how many of the reports probably had an astronomical basis.

But the reported actions of the flying saucers did not fit the expected pattern of advanced technological military devices, and only a fraction could with certainty be ascribed to astronomical objects or events. Opinion in Project Sign soon became markedly divided: was it foreign technology or *really foreign* technology? Craft from outer space? A public psychosis? A fad spawned by postwar nerves?

The division grew greater as it became increasingly clear that the "ordinary" foreign technology explanation was untenable. An "explanation gap" had arisen. Either the whole phenomenon had to be *psychological* (an expression that was often used for want of a cogent explanation), or there was something behind the phenomenon that no one wanted to admit. When the mind is suddenly confronted with "facts" that are decidedly uncomfortable, that refuse to fit into the standard recognized world picture, a frantic effort is made to bridge that gap emotionally rather than intellectually (which would require an honest admission of the inadequacy of our knowledge).

Frenetic efforts are made either to contrive an ad hoc explanation to save the phenomenon or to discredit the data. When we are faced with a situation that is well above our threshold of acceptability, there seems to be a built in mental censor that tends to block or to sidestep a phenomenon that is too strange and to take refuge in the familiar.

The history of science is replete with "explainings away" in order to preserve the status quo. Discovery of fossils of extinct species, pointing strongly to the concept of biological evolution, was met with many contrived attempts to demolish the fossil fingers pointing unmistakably to Darwinian evolution. Many, too, were the pat explanations before facts finally demanded the acceptance of the theory of circulation of the blood, the heliocentric hypothesis, hypnotism, meteorites, disease-causing bacteria, and many other phenomena that are accepted today.

In 1948 Project Sign faced a major explanation gap and sought the aid of its scientific advisers, both in the Air Force and in the scientific fraternity. Their reaction was the expected one, the one that has been experienced through the centuries: "It can't be; therefore, it isn't." The explanation gap was far above the threshold of acceptance, so the expected refusal to weigh and consider, the popularly accepted hallmark of the scientist, came to the fore.

In fairness to the scientific fraternity it must be emphasized that available data were poorly presented and were mixed with substantial quantities of nonsense—stupid reports, misperceptions of Venus and meteors by the untutored—all emotionally charged.

In my restricted assignment with Project Sign (merely to weed out reports ascribable to astronomical stimuli) I soon became aware of cases that, taken at face value, were outstanding challenges to science. But could they be so regarded? It was clear to me that because of the paucity of hard-core scientific data in the reports, their mystery might easily evaporate if such reports were properly followed up and investigated. The problem of the lack of proper investigation was present from the start.

As a junior in the ranks of science at that time, and not inclined to be a martyr or to make a fool of myself on the basis of incomplete data, I decided to remain neutral and let the phenomenon prove or disprove itself. Senior advisers to the Pentagon had shown themselves universally scornful of the flying saucer problem, and I had to admit to myself that although the data were provocative, they fell far short of being scientifically conclusive. It was not until several years had passed and data of similar nature continued to flow not only from this country but from many others that I had occasion to feel that the phenomenon was indeed being proved: there were too many occurrences that couldn't be explained in ordinary terms.

The Pentagon's official attitude was largely dictated by the scientific fraternity. After all, not even a major general wishes to be laughed at by highly placed members in the scientific hierarchy. One example was, of course, Harvard astronomy professor Dr. Menzel, who took a seemingly compulsive interest in the flying saucer question, even though this subject was far removed from his scientific field. He loudly proclaimed that UFOs were nonsense and particularly championed the "mirage theory" of flying saucers. He ascribed properties to mirages, and mirage properties to UFOs, which have since been shown to be completely untenable, even by the Air Force itself.[2]

But we must not single out Dr. Menzel for succumbing to the "explanation gap" syndrome, although he undoubtedly helped influence the official Pentagon "scientific" position. Nearly all scientists gave short shrift to the problem, some taking great delight in pontificating before the lay public. Much of the opprobrium of science was well founded. Reports based on misperceptions abounded, and the crackpot contingent was always lurking in the wings or making its presence known through flying saucer cult movements and pseudo-religious tracts and pronouncements.

I confess much pleasure in providing discomfiture to such groups or to misguided and excitable UFO report generators. It was satisfying, for instance, to establish that one report of a "mother ship and four companions" unquestionably arose from observation through a small telescope of the planet Jupiter and its four major satellites and to prove that what one woman had called a "spaceship with tail lights" was nothing more than a bright meteor.

But the 237 original Project Sign reports were not convincing and did not support "visitors from space." In reviewing these cases again in 1970, I recognized their markedly poor quality. Reports in the 1950s and in 1966 and 1967 were of a much higher caliber in both Strangeness Rating and in what had been determined about the character of the witnesses. In 1947–1948 there really wasn't too much to get excited about. There were certainly some reports that if taken at near-face value, suggested no possible normal physical explanation, but even these were poor in that they had been inadequately investigated; many crucial data were missing. Among the first 237 UFO cases there were no Close Encounters that approached the quality of the details of more recent reports in that category, and there were only a few (poor) radar reports. Daylight Discs were the preponderant category in the puzzling Project Sign cases, and there were only a few provocative Nocturnal Light cases.

Even today, if I were given only the data of these first Air Force cases without knowing the UFO report patterns that later became evident all over the world, I would still come to the conclusion I reached in 1949: there were a number of truly puzzling reports for which the data were not sufficient to base positive conclusions. Even so, I would repeat my conclusion of 1953: the subject is worthy of further scientific investigation.

In all fairness to the Air Force, we must remember that as much as it has been justifiably maligned for its treatment of UFOs, its mission, particularly in 1949, was not one of science but of defense. The Air Force's responsibility was discharged when they demonstrated that the UFO phenomenon showed no immediate evidence of being hostile and was not a threat to our national security.

If at that point the Air Force had turned the problem over to a recognized and long-established nonprofit scientific organization, the history of the UFO problem might well have been quite different. A small scientific task force of persons with an understanding of the basic problems, set up on a continuing basis to publish their findings in recognized journals from time to time, would have sufficed.

Instead, the Air Force adopted another path. Once the Pentagon had set firm UFO policy and had rejected the historic "Estimate of the Situation" (which one faction in Project Sign had sent through channels clear to the top), in which it was concluded that flying saucer reports did give evidence of extraterrestrial intelligence, the Air Force entered upon a long period of unfortunate, amateurish public relations. The issuance of propaganda and public relations handouts, which were often ill-considered and contradictory, ushered in an era of confusion from 1950–1970. The insistence on official secrecy and frequent classification of documents was hardly needed since the Pentagon had declared that the problem really didn't exist.

The role of the Air Force during this era was both pivotal and enigmatic. It was pivotal because the world (specifically other governments the offices of which were also the recipients of UFO reports) took its cue from the US Air Force. When I asked what was being done in those countries about the UFO problem, on many occasions I was told that since the United States, with all its funds and facilities, was handling the problem, what more could other countries do with their limited facilities? They would await the outcome of the US investigation.

It was enigmatic because of the obvious question: if there was nothing whatever to the UFO phenomenon other than misperceptions, hoaxes, etc., why continue a UFO program? Why adopt a confusing and misleading public relations posture that on many occasions led to insulting the intelligence of competent people? Some of the Blue Book evaluations of sincere reports were often so transparent and irrelevant that they had later to be retracted. Was this all a smoke screen, a cover-up job for which Project Blue Book was a front, the real work and information being handled by another agency?

Had there been initiated at an early stage a continuing scientific commission or institute, both scientific respectability for the subject and a dignified image for the Air Force in this area would have been gained. The public could have been made aware, through nonsensational channels, of what was puzzling and not yet known, which reports had been verifiably demonstrated to have been the results of misperceptions, hoaxes, etc. Since the public no longer would have had to fear ridicule, its assistance and interest would have been assured. It might have been determined whether there was any "signal" in the "noise"; whether there was, in the global UFO reports, genuinely new empirical material. A nonmilitary scientific commission operating in a dignified key would probably have had the cooperation of international scientific groups. It was a mistake from the start to shroud the subject in an air of military science fiction, an error compounded further by seeming duplicity in public pronouncements. If the quietly working scientific group had disclosed, after dedicated study, that there were no indications of scientific paydirt in the reports, this would have been generally accepted, the cultists and crackpots, of course, excepted. Conversely, detailed studies and research in depth could have been instituted had there been such indications.

The course that was followed was quite different. Project Sign started with a 2A priority, 1A being the highest. Shortly after becoming consultant to the project, I learned that there had been internal dissension from the start. There had been those who insisted that flying saucers were Russian devices; others thought that they were from outer space, and still others, of course, thought that the subject was entire nonsense.

On my assignment I was asked to work entirely independently of the other consultants and Project Sign members. This separation apparently was to ensure that I would remain unbiased. My final report, compiled with the able assistance of Mrs. Charles Sumerson, was issued after Project Sign has somewhat mysteriously been transformed into Project Grudge, on February 11, 1949. I was not aware of the change as I continued to do my best to find logical astronomical explanations for as many of the 237 reports as possible.

The change to Project Grudge signaled the adoption of the strict brush-off attitude to the UFO problem. Now the public relations statements on specific UFO cases bore little resemblance to the facts of the case. If a case contained some of the elements possibly attributable to aircraft, a balloon, etc., it automatically became that object in the press release.

Captain Ruppelt, speaking of these brush-offs as part of an intentional smoke-screen to cover up facts by adding confusion, wrote, "This is not true; there was merely a lack of coordination. But had the Air Force tried to throw up a screen of confusion, they couldn't have done a better job."[3] As an example Ruppelt quotes from a Pentagon news release that indicated that flying saucers were (a) metoric breakup such that their crystals cast the light of the sun, (b) sunlight on lowhanging clouds, and (c) hailstones that became flattened out and glided. Ruppelt was right when he said, "The problem was tackled with organized confusion." Confidence in the Air Force's ability or willingness to cope with this problem was ebbing as early as 1949. Ruppelt has characterized this period most excellently in his Report on Unidentified Flying Objects. Indeed, his book should be required reading for anyone seriously interested in the history of this subject. In my contacts with him I found him to be honest and seriously puzzled about the whole phenomenon.

The transition from Project Sign to Project Grudge came before my report was issued, and by the time I submitted my report, the climate toward any serious investigation of flying saucers had become very chilly. "This drastic change in official attitude," wrote Ruppelt, "is as difficult to explain as it was difficult for many people who knew what was going on inside Project Sign to believe."[4] He also wrote, "This period of 'mind-changing' bothered me. Here were people deciding that there was nothing to this UFO business right at the time when the reports seemed to be getting better. From what I could see, if there were any mind-changing to be done, it should have been the other way. . . ."[5]

I can fully support this opinion. The earliest reports, particularly those I first studied in Project Sign, were of very much poorer quality than those that began to come in later. Some were limited to a few dozen words, with details necessary for adequate evaluation missing.

Ruppelt ascribed the change in attitude to the fact that the military wants answers, not mysteries. "Before, if an interesting report came in," he writes, "and they wanted an answer, all they'd get was an, 'it could be real, but we can't prove it'. Now such a request got a quick snappy, 'It was a balloon,' and feathers were stuck in the caps from ATIC all the way up to the Pentagon. Everybody felt fine."[6]

Ruppelt described the period following the start of Project Grudge as the Dark Ages. New personnel, rather than the most experienced people in Project Sign, established and used the Air Force theorem: "It can't be; therefore, it isn't." Ruppelt says, "Everything was being evaluated on the premise that UFOs couldn't exist,"[7] and, "Good UFO reports continued to come in at the rate of about ten per month, but they weren't being verified or investigated. Most of them were being discarded."[8]

In the years that followed, when I was consultant to Project Blue Book, no report that came in through official military channels was discarded, but only the most perfuctory attempts were made to mount any type of serious investigation. This was especially true of the particularly puzzling, unusual cases. These were frequently evaluated as Unidentified and put aside. The objective had been attained: the UFO had been identified as Unidentified.

After I submitted my report, in April, 1949; shortly after Project Grudge was underway, I was completely severed from the UFO office in Dayton. Thus I did not know until later what went on during the Dark Ages.

My report itself ran to better than 300 pages, many of them nearly blank, for all the page contained was the statement, "There is no astronomical explanation for this report." My obligation was discharged. Sometimes I ventured further: "We can conjecture that a cluster of balloons (cosmic ray apparatus) was observed, the motion of which was merely the reflection of the motion of the plane."

In the introduction to the report I wrote, "Among the general public, two attitudes towards flying saucers seem to be prevalent: one, that all sightings are misidentifications or hoaxes, and two, 'that there must be something to it.' From the outset, I have attempted to regard each report . . . as an honest statement by the observer and to adhere to neither of the two attitudes."

I noted what was then plaguing and what was to continue to plague the UFO office: the incompleteness of the data and of any effort to upgrade it. "Almost all of the data dealt with in this 300-page report are incomplete and inexact, and some are distinctly contradictory. Therefore, it has obviously been impossible to reach definite scientific conclusions. Most conclusions are offered in terms of probability, the degree of which is discussed in the individual reports."

Some two months earlier Project Sign in a secret report, which I did not see until years later, stated:

> No definite evidence is yet available to confirm or disprove the actual existence of unidentified flying objects as new and unknown types of aircraft. A limited number of the incidents has been identified as known objects.
>
> Based on the possibility that the objects are really unidentified and unconventional types of aircraft, a technical analysis is made of some of the reports to determine the aerodynamic, propulsion, and control features that would be required for the object to perform as

> described in the reports. The objects sighted have been grouped into four classifications according to configurations:
>
> 1. Flying discs, i.e., very low aspect ratio aircraft,
> 2. Torpedo or cigar-shaped bodies with no wings or fins visible in flight,
> 3. Spherical or balloon-shaped objects,
> 4. Balls of light.
>
> The first three groups are capable of flight by aerodynamic or aerostatic means and can be propelled and controlled by methods known to aeronautical designers.

Even in 1949 the UFOs came in the same patterns, which persisted for the ensuing years.

The "frustration barrier" continued. No real attempt was ever made to gather all the data that were available. The Air Force investigators had not bothered to gather what was there. In many instances, starting from a mere item on the back pages of a small town newspaper, I have been able to reconstruct, with the patient aid of the observers, a coherent account of reported events, and generally I have found the persons concerned fully cooperative once they were assured that no ridicule or unfavorable publicity would result from the interview. Blue Book files are replete with cases labeled Insufficient Information, whereas in many cases the proper label should have been, Insufficient Follow-up.

It became patently clear to me as the years passed that no Blue Book case had been given the "FBI treatment"; that is, no case was followed through until every possible clue or bit of evidence was obtained, as is standard procedure in kidnapping, narcotics rings, and bank robbery cases.

Quite the opposite attitude was taken by Blue Book. When a case did appear to have a likely misperception explanation (and hence should have been excluded from further UFO investigative effort), Blue Book often spared little effort in phone calls, interrogations, etc., in order to pin it down to a planet, a refueling mission, or some other natural occurrence. Thus they set their dogs to catching simple chicken thieves but ignored potentially far more important prey.

Had there been available, for the many hundreds of Blue Book cases now carried as Unidentified, a scientifically trained and conscientious investigator with immediate reaction capability (immediate access to transportation to the locale of the reported event within twenty-four hours) far more information would have been gathered. The true Strangeness Rating and Probability Rating for each case could have been determined with some confidence. I had made several attempts, including some before Congressional subcommittee, for such immediate reaction capability—but to no avail.

Often Blue Book did not bother to investigate until the UFO event had attained some prominence in the press (the Portage County case was a good example) or until an inquiry was made from a Congressperson whose constituents felt they had not been treated right by Blue Book. Nothing brought more immediate and frenzied reaction from Blue Book than a query from Congress. Then, however, the effort was directed to the composing of a quick but satisfactory answer rather than to a serious study of the case. I frequently observed occasions when the sole Blue Book objective was "getting the Congressperson off its back" by constructing some sort of possible explanation rather than mounting a scientific investigative effort.

Thus the program did not change through the years. Reports came in and were handled in a completely routine manner, always on the assumption that they had been spawned by untutored people unable to identify perfectly natural occurrences. When the going really got tough, the label "unidentified" was used, but the investigative effort ended there. It was tacitly assumed that *had* an exhaustive effort been made to identify the source of the report, it would not have been successful. Why, then, if we can assume that, should any detailed effort be wasted on such an Unidentified phenomenon?

Through the years, the percentage of Unidentifieds remained essentially the same. Table 1 covers the first 237 UFO reports received by the Air Force; it shows that some 20 percent of these met the present definition of UFO, that is, they stumped the experts. Twenty years later the Condon committee, using presumably a better selection of reports and more scientists, were unable to find solutions for more than 25 percent of the cases they examined. Through the years there seems to have been a stubborn, unyielding residue of "incredible reports from credible people."

On October 7, 1968, I addressed my reply to Colonel Sleeper:

> I address my report to you alone, for as will be apparent, should the present staff of Blue Book read it, any further personal contact with them would prove most embarrassing to all parties concerned. . . . It may be of interest to you that, in all of my twenty years as consultant, you are the first commander who has ever asked me to write an evaluation of Blue Book. I would have been happy to do so earlier, but on those occasions when I attempted to advise on procedures and methodology . . . I had been politely but firmly reminded . . . of my place in the organization.
>
> I sincerely hope that at long last I may help transform Blue Book into what the public and the scientific world has been told it is . . . an investigative organization dedicated to the defense of the country but doing a good scientific job also. . . . It is time that Blue Book no longer be called, as some wag has done, "the Society for the Explanation of the Uninvestigated."

Table 1

EVALUATIONS BY J. ALLEN HYNEK IN 1948–1949 OF THE FIRST 237 UFO REPORTS RECEIVED BY THE AIR FORCE

	NO. OF INCIDENTS	APPROX. %
1. Astronomical		
a. High probability	42	18
b. Fair or low probability	33	14
	75	32
2. Non-astronomical but suggestive of other explanations		
a. Balloons or ordinary aircraft	48	20
b. Rockets, flares, or falling bodies	23	10
c. Miscellaneous (reflections, auroral streamers, birds, etc.)	13	5
	84	35
3. Non-astronomical, with no evident explanation		
a. Lack of evidence precludes explanation	30	13
b. Evidence offered suggests no explanation	48	20
	78	33

You have chosen to refer to methods of "product improvement." Although this is a metaphor scientists rarely use, I believe it is a happy one in that it is practical to think in terms of what the Blue Book product is, who the consumer . . . is, how the product is "packaged," what the product "image" is, and how we might "tool up" for product improvement.

You have indicated that I should not concern myself with history of Blue Book; however . . . the 1960 hearings in Washington are germane to this report. My recommendations at that time for changes in Blue Book were applauded by the Smart Committee but never funded (although funds were promised), so my efforts came to naught.

Since my report is rather long, I have prefaced it with a sequential summary of points covered and of recommendations made.

Summary

A. It is concluded that neither of the two missions of Blue Book (AFR 80-17), (1) to determine if the UFO is a possible threat to the United States and (2) to use the scientific or technical data gained from a study of UFO reports, are being adequately executed.

B. The staff of Blue Book, both in numbers and in scientific training, is grossly inadequate to perform tasks assigned under AFR 80-17.

C. Blue Book suffers intramurally in that it is a closed system that has fallen victim to the closed loop type of operation. There has been virtually no scientific dialogue between Blue Book and the outside scientific world. Totally inadequate use is made of the extensive scientific facilities of the Air Force in executing the Blue Book mission. The superb talents and facilities of AFCRL [Air Force Cambridge Research Laboratories] and of AFOSR [Air Force Office of Scientific Research], for instance, have rarely been used. The lack of scientific dialogue between members of Blue Book and outside scientists has been appalling.

D. The statistical methods employed by Blue Book are nothing less than a travesty.

E. There has been a lack of attention to significant UFO cases, as judged by this consultant and others, and too much time spent on routine cases that contain few information bits and on peripheral public relations tasks. Concentration should be on two or three potentially scientifically significant cases per month, rather than having Blue Book effort spread thin over forty to seventy cases per month. Too much attention has been paid to one-witness cases and to cases in which only point-source lights on the sky are seen at night and far too little to the cases of high Strangeness Rating reported by witnesses of conceded reputation.

F. The information input to Blue Book is grossly inadequate. An impossible load is placed on Blue Book by the almost consistent failure of UFO officers at local air bases to transmit adequate information to Blue Book. Many information bits that could have been obtained by conscientious interrogation by the UFO officer are omitted, throwing the burden upon Blue Book to reopen interrogation for additional information, sometimes of the most elementary but necessary sort—e.g., wind directions, angular sizes and speed, details of trajectory, qualifications and nature of witnesses, additional witnesses, etc. The upgrading of original data is the most pressing need within Blue Book.

G. The basic attitude and approach within Blue Book is illogical and unscientific in that a working hypothesis has been adopted which colors and determines the method of investigation. One might put it in the form of a *Theorem*:

   For any given reported UFO case, if taken by itself and without respect and regard to correlations with other UFO cases in this and other countries, it is always possible to adduce a possible even though far-fetched natural explanation, if one operates solely on the hypothesis that all UFO reports, by the very nature of things, must result from purely well known and accepted causes.

   The theorem has a *Corollary*:

   It is impossible for Blue Book to evaluate a UFO report as anything other than a misidentification of a natural object or phenomenon, a hoax, or a hallucination. (In those relatively few cases where even this procedure met with difficulty, the report was evaluated as "Unidentified" but with no indication that the theorem had been outraged.)

H. Inadequate use has been made of the Project scientific consultant. Only cases that the *project monitor* deems worthwhile are brought to his attention. His scope of operation, including personal direct access to both unclassified and classified files, has been consistently limited and thwarted. He often learns of interesting cases only a month or two after the receipt of the report at Blue Book, and no attempt is made to bring the consultant into the operating loop except in the most peripheral manner.

The popular impression through the years was that Blue Book was a full-fledged, serious operation. The public perhaps envisioned a spacious, well-staffed office with rows of file cabinets, a computer terminal for querying the UFO data bank, and groups of scientists quietly studying reports, attended by a staff of assistants.

The actual situation was unfortunately the opposite. The operation was generally headed by an officer of lesser rank. In the military the importance attached to a mission is usually in direct proportion to the rank of the commanding officer. The relatively low-ranking officers in charge of Blue Book were usually assisted by a lieutenant and sometimes only by a sergeant. For one long period of time a sergeant with little technical training was given the chore of evaluating most of the incoming reports.

This was not exactly a first-line, high priority operation. Blue Book had much too small a staff to do justice to a phenomenon that so often greatly concerned the public. Compounding the problem, the staff was able to devote only part of its time to the technical problem at hand. During my regular visits to Blue Book across the years I observed that much of the work in the office was devoted to peripheral matters, all done at a leisurely pace.

Further, Blue Book's low-ranking officers had no leverage to initiate the type of investigations that were needed and for which I frequently asked. The military is entirely hierarchical; a captain cannot command a colonel or a major at another base to obtain information for him. He can only request.

As long as Blue Book did not have at least a full colonel in command, it was impossible to execute its assigned task properly. In reviewing cases that had come in during the previous month, I often asked that additional, often crucial information on a case be obtained. The results were at best minimal; officers at other bases were generally too busy to bother to investigate further. Why should they? They all knew it was a finger exercise anyway.

Blue Book was a cover-up to the extent that the assigned problem was glossed over for one reason or another. In my many years association with Blue Book, I do not recall even one serious discussion of methodology, of improving the process of data gathering or of techniques of comprehensive interrogation of witnesses.

The reader may well ask at this point why I did not either lay seige to the Pentagon, demanding action, or simply resign in disgust. Temperamentally, I am one who can easily bide his time. I also dislike a fight, especially with the military. But most importantly, Blue Book had the store of data (as poor as they were), and my association with it gave me access to those data. In a sense I played Kepler to Blue Book's Tycho Brahe.[9]

As far as demanding action from the Pentagon, I knew only too well the prevailing climate and recognized that had I been too outspoken, I would have quickly been discredited, labeled a UFO nut, lost access to data, and certainly would have lost all further effectiveness. I have always been of the turn of mind that "trust will out" if given time; if there was indeed scientific paydirt in the UFO phenomenon, as time went on and the gathering of data improved, even the most hostile skeptics would be powerless to sweep it under the carpet. The astronomer traditionally adopts a very longtime scale.

By and large, however, Blue Book data were poor in content, and even worse, they were maintained in virtually unusable form. With access to modern electronic data processing techniques, Blue Book maintained its data entirely unprocessed. Cases were filed by date alone, and not even a rudimentary cross-indexing was attempted. Had the data been put in machine readable form, the computer could have been used to seek patterns in the reports, to compare the elements of one report with those of another, and to delineate, for instance, the six basic categories of sightings used in this book. Since all the thousands of cases were recorded only chronologically, even so simple a matter as tabulating sightings from different geographical locations, from different types of witnesses, etc., was impossible except by going through, manually, each and every report. A proposal[10] for elementary computerization of the data in the Blue Book files, devised by Jacques Vallee and myself and submitted by me directly to Major Quintanilla at Blue Book, was summarily turned down.

In view of the above and of the frequently contradictory and inane public relations statements concerning UFO reports, which even the man on the street found unconvincing, it is hardly a wonder that the charge was frequently made that the publicly visible Air Force "investigation" of UFOs was merely a front for a real investigation being carried on somewhere higher up.

Were I the captain of a debating team whose job it is, of course, to marshal all the facts favorable to his side and studiously to avoid the other's, I could defend either side of the argument. At no time, however, did I encounter any evidence that could be presented as valid proof that Blue Book was indeed a cover-up operation. However, many indications, bits of information, and scraps of conversation could be force-fitted into a "yes" for the cover-up thesis. Thus, for instance, one time when I inquired into the specifics of a certain case, I was told by the Pentagon's chief scientist that he had been advised by those at a much higher level to tell me "not to pursue the matter further." One can make of that what one will.

In a country as security conscious as is ours, where central intelligence is a fine art, it frequently seemed to me that very provocative UFO reports were dismissed without any seeming follow-up—certainly an illogical if not dangerous procedure unless one knew a priori that the report really was of no potential information value to the security of the country (or that it was but was being taken care of elsewhere). As an example, the report of five rapidly moving discs, made by a member in good standing of the 524th Intelligence Squadron stationed in Saigon and observed by him from the roof of the squadron's headquarters, went untouched by Major Quintinalla and Blue Book on the grounds that "the sighting was not within the continental limits of the United States." It would seem almost inconceivable that the intelligence officer in question would not have been further interrogated by some agency; certainly in an active battle area his sighting might have presaged a new military device of the enemy.

Another example, one of many, was this: on the first day of August, 1965, and on the following two days there occurred the "Midwest flap." From several states strange Nocturnal Lights were reported by ostensibly reliable police officers on patrol at various places over an area of several hundred square miles. Blue Book dismissed this event as "stars seen through inversion layers," although I know of no astronomer who has ever witnessed inversion effects that produced these reported effects. Both past experience and calculations show that such illusory effects, in which stars move over at a considerable arc of the sky, simply cannot be produced by thermal inversions.

However, police officers weren't the only ones to report. The following is a direct transcript of a Blue Book memo: In the early morning hours of August 1, 1965, the following calls were received at the Blue Book offices by Lieutenant Anspaugh, who was on duty that night:

> 1:30 A.M.—Captain Snelling, of the US Air Force command post near Cheyenne, Wyoming, called to say that fifteen to twenty phone calls had been received at the local radio station about a large circular object emitting several colors but no sound, sighted over the city. Two officers and one airman controller at the base reported that after being sighted directly over base operations, the object had begun to move rapidly to the northeast.
>
> 2:20 A.M.—Colonel Johnson, base commander of Francis E. Warren Air Force Base, near Cheyenne, Wyoming, called Dayton to say that the commanding officer of the Sioux Army Depot saw five objects at 1:45 A.M. and reported an alleged configuration of two UFOs previously reported over E Site. At 1:49 A.M. members of E flight reportedly saw what appeared to be the same uniform reported at 1:48 A.M. by G flight. Two security teams were dispatched from E flight to investigate.
>
> 2:50 A.M.—Nine more UFOs were sighted, and at 3:35 A.M. Colonel Williams, commanding officer of the Sioux Army Depot, at Sydney, Nebraska, reported five UFOs going east.
>
> 4:05 A.M.—Colonel Johnson made another phone call to Dayton to say that at 4:00 A.M. Q flight reported nine UFOs in sight: four to the northwest, three to the northeast, and two over Cheyenne.

> 4:40 A.M.—Captain Howell, Air Force Command Post, called Dayton and Defense Intelligence Agency to report that a Strategic Air Command Team at Site H-2 at 3:00 A.M. reported a white oval UFO directly overhead. Later Strategic Air Command Post passed the following: Francis E. Warren Air Force Base reports (Site B-4 3:17 A.M.)—A UFO 90 miles east of Cheyenne at a high rate of speed and descending—oval and white with white lines on its sides and a flashing red light in its center moving east; reported to have landed 10 miles east of the site.
>
> 3:20 A.M.—Seven UFOs reported east of the site.
>
> 3:25 A.M.—E Site reported six UFOs stacked vertically.
>
> 3:27 A.M.—G-1 reported one ascending and at the same time, E-2 reported two additional UFOs had joined the seven for a total of nine.
>
> 3:28 A.M.—G-1 reported a UFO descending further, going east.
>
> 3:32 A.M.—The same site has a UFO climbing and leveling off.
>
> 3:40 A.M.—G Site reported one UFO at 70° azimuth and one at 120°. Three now came from the east, stacked vertically, passed through the other two, with all five heading west.

When I asked Major Quintanilla what was being done about investigating these reports; he said that the sightings were nothing but stars! This is certainly tantamount to saying that our Strategic Air Command, responsible for the defense of the country against major attacks from the air, was staffed by a notable set of incompetents who mistook twinkling stars for strange craft. These are the people who someday might have the responsibility for waging a nuclear war.

For some, incidents such as the above would be prima facie and conclusive evidence that the cover-up hypothesis was the correct one, on the grounds that no group charged with serious defense responsibilities for the country could have been so stupid.

On the other hand, our hypothetical debating team captain could amass an even more impressive cache of evidence to conclude quite the opposite: that the entire Blue Book operation was a foul-up based on the categorical premise that the incredible things reported could not possibly have any basis in fact. After all, science pretty well understands the physical world and knows what's possible and what is not. Since the reported actions of UFOs clearly didn't fit this world picture, they simply *had to be* figments of the imagination produced in one way or another.

All my association with Blue Book showed clearly that the project rarely exhibited any scientific interest in the UFO problem. They certainly did not address themselves to what should have been considered the central problem of the UFO phenomenon: is there an as yet unknown physical or psychological or even paranormal process that gives rise to those UFO reports that survive severe screening and still remain truly puzzling?

Such lack of interest belies any charge of cover-up; they just didn't care. There is another argument for the noncoverup viewpoint: the underlings in the military hierarchy (and all Blue Book officers were such—generally captains or majors, two of which finally made lieutenant colonel but never full colonel) looked mainly toward two things: promotion and early retirement. Therefore, in controversial issues it was always considered far wiser not to rock the boat, to please the superior officer rather than to make waves. Thus, when the superior officers, who did not know the facts but were wedded to a rigid framework of military thinking handed down from above, let it be known in any controversial issue (whether UFOs or not) what the "right way" of thinking is, no underling officer was going to oppose or even question it unless, of course, he was 99 percent certain that he could prove himself correct in the controversy—and quickly.

Since the Pentagon had spoken in no uncertain terms about UFOs, no Blue Book officer in his right promotion-conscious military mind was going to buck that, even if he had private opinions on the matter.

Another factor added to the noncover-up theory. Turnover in the Blue Book office was rather high. Sooner or later the officer in charge would be out of it, just that much closer to promotion and retirement, if he just sat tight. From 1952 to 1969 the office was headed in turn by Captain Ruppelt (who did not make his own views known until he was out of the Air Force), Captain Hardin (who had ambitions to be a stock broker), Captain Gregory (to whom promotion was the be-all and end-all of existence), Major Friend, and finally Major Quintanilla, who had the longest term of office. Of all the officers I served with in Blue Book, Colonel Friend earned my respect.

Whatever private views he might have held, he was a total and practical realist, and sitting where he could see the scoreboard, he recognized the limitations of his office but conducted himself with dignity and a total lack of the bombast that characterized several of the other Blue Book heads.

Thus one can have one's choice of whether Blue Book was a front or merely a foul-up. But that there was certainly foul-up and complete divorce from the scientific community within Blue Book was apparent. The members of the scientific fraternity were, of course, wedded to the misperception-delusion hypothesis (there was no need for interchange of ideas with Blue Book, which held the same views), and some members rose to heights of vitriolic verbiage in denouncing reporters of UFOs. This phase of the total phenomenon had many of the aspects of a modem witch-hunt.

But Blue Book is no more, and its closing raised the question: to whom does one report current UFO sightings? That such sightings continue to occur (and to be reported unofficially) is incontrovertible, as is shown by any news-clipping service that covers small-town newspapers and out-of-the way publications. At the time of this writing there is no government or officially designated agency to whom such a report can be made.[11]

There are many UFO organizations throughout the world that avidly accept UFO reports, often far too avidly and uncritically, in order to have material for their publications. During the past score of years literally hundreds of civilian UFO organizations began in many parts of the world, particularly in France, England, Germany, Japan, Italy, Australia, some of the Latin American countries, and, of course, the United States. Many of these were short-lived, but each in its way was the recipient of UFO reports covering a wide spectrum of reliability and credibility. Most organizations received reports and did little screening or serious investigating. Generally this stemmed not from a lack of interest or even of ability (although some groups were innocent of the rudiments of scientific procedures) but from lack of funds and time.

Many of the organizations published bulletins on a more-or-less scheduled basis. Often these were just mimeographed sheets, but most of the publications were even more short-lived than their parent organizations. A few journals existed and still do, independently of private sponsoring investigative groups. Outstanding among these is *FSR, Flying Saucer Review*, published in London since 1954. It is a veritable treasure house of UFO reports, some of which have been investigated rather thoroughly but the majority of which rank with the average Blue Book report. The reader suffers a strong sense of frustration in reading such reports; in each case the conscientious reader longs for more details, but only rarely are they given. Unfortunately there is no journal that is financially supported sufficiently to devote pages to details for the relatively few serious investigators. The existing journals have subscribers who, for the most part, are satisfied with summaries; indeed, I am afraid some of their readers wish merely to be titillated by incredible tales.

France has two particularly outstanding publications, *Phenomenes Spatiaux* and *Lumières dans la Nuit*. They serve as publication organs for dedicated groups of investigators—mainly in France—who quietly have gone about collecting data of good quality in perhaps a more systematic way than groups in other parts of the world. Such groups give freely of their time and limited funds in painstakingly tracking down UFO witnesses and conducting able interrogations. Australia, New Zealand, Japan, Canada, Sweden, and Italy are some of the other countries in which UFO journals or bulletins are published. There is a need for an international organization that might act as a clearing house for such journals and their contents. It has been suggested on a number of occasions that the United Nations, perhaps UNESCO, might act in such a capacity, but to date all such suggestions have been tabled.

In the United States there have been only two viable civilian UFO investigative organizations. The older of these, APRO (Aerial Phenomena Research Organization), now located in Tucson, Arizona, was started in 1952 in Wisconsin. APRO has done an excellent job in collecting UFO data, resumés of which have been published in the *APRO Bulletin*.

Four years after APRO was organized, NICAP (National Investigating Committee for Aerial Phenomena), located in Washington, DC, was formed and grew to have the larger membership. Both organizations have suffered from lack of financial support, which meant, of course, that costly investigations were not possible. APRO has an international outlook, as is evidenced by the sizable number of foreign consultants and associates.

Although both organizations are serious-intentioned collectors of UFO data, each unavoidably numbers in its membership overenthusiastic and uncritical persons enamored of the idea of UFOs. Nonetheless, neither APRO nor NICAP are in any sense of the word "crackpot" organizations and have many serious members, many of whom have considerable technical and scientific training.

There was very little overlap in reports between Blue Book and either APRO or NICAP. Dr. Saunders had remarked that in compiling reports for the abortive Condon committee computer program, the only overlap in reports occurred for the well-publicized cases. Otherwise, the three organizations had essentially independent UFO files.

Now that Blue Book is no more, I am often asked whether the Air Force is really out of the UFO business. The answer is probably contained in an official letter from the Pentagon, written after the close of Blue Blook. It states:

> The Aerospace Defense Command (ADC) is charged with the responsibility for aerospace defense of the United States. . . . Consequently, ADC is responsible for unknown aerial phenomena reported in any manner, and the provisions of joint Army-Navy-Air Force publication (JANAP-146) provide for the processing of reports received from nonmilitary sources.

In JANAP-146 E, currently in force, provisions and instructions for the reporting of unknown objects in the air by military personnel are explicitly set forth. It must be remembered, however, that the military is primarily interested in unidentified planes, especially those that might be foreign. Such planes certainly are unidentified flying objects, although they do not satisfy the definition used in this book. There never is any question but that they are planes (which are flying objects) and that their source is unidentified.

In my long association with Blue Book, I had some very interesting encounters with UFO witnesses, some equally interesting but less enjoyable encounters with military personnel, and a most intimate view of the running of a pseudoscientific project. Blue Book was essentially a closed operation, in which A talked to B, and B talked to C, and C talked to A. There was little input from outside scientific groups. It is conceivable that in its ingrown and official military way, it was allowed to bumble along, while apart from it serious official attention was given to a few selected cases that might not even have gone through Blue Book channels. I simply do not know. In my position as periodic consultant, I certainly was never taken into the confidence of higher Pentagon officials concerning these matters. JANAP-146E still exists and is in effect, and it calls for the processing of reports of unknown aerial phenomena from both military and nonmilitary sources. Probably little more need be said.

CHAPTER TWELVE

Science Is Not Always What Scientists Do

It is the duty of Science, not to discard facts merely because they seem to be extraordinary and that it remains unable to explain them.

—ATTRIBUTED TO ALEXIS CARREL

On October 6, 1966, the University of Colorado and the US Air Force entered into formal agreement to establish a scientific committee to study (and presumably to settle once and for all) the vexing problem of UFOs with which the Air Force had been saddled for twenty years. It was to be directed by Dr. Edward U. Condon, a physicist of established reputation, noted not only for his scientific record but for his courage in speaking out on controversial issues.

Two years later there appeared the results of the committee's work: a voluminous, rambling, poorly organized report of 937 pages of text, considerably less than half of which was addressed to the investigation of UFO reports. The report opened with a singularly slanted summary by Dr. Condon, which adroitly avoided mentioning that there was embodied within the bowels of the report a remaining mystery; the committee had been unable to furnish adequate explanations for more than a quarter of the cases examined.

However, were it not for the fact that the public has had ready access through the press only to the summary of the report, and its implication that the UFO problem has been "solved," there would be little point to this critique of the Condon Report. The report covers little new territory. Others before Condon had demonstrated that the data at hand were far from sufficient to establish the hypothesis of extraterrestrial visitation. Condon only partially retraced the steps of those more knowledgeable than he and his group.

Conclusions and recommendations comprise the first part of the two-chapter summary. Two statements are particularly illuminating:

> Careful consideration of the record as it is available to us leads us to conclude that further extensive study of UFOs probably cannot be justified in the expectation that science will be advanced thereby.

This was surely the kiss of death to any further investigation in the name of the quest for knowledge. Yet in a somewhat platitudinous vein we find the statement:

> Therefore we think that all of the agencies of the federal government, and the private foundations as well, ought to be willing to consider UFO research proposals along with the others submitted to them on an open-minded, unprejudiced basis. While we do not think at present that anything worth-while is likely to come of such research, each individual case ought to be carefully considered on its own merits.

Truly a masterpiece of throwing a scrap of political meat to the critic dogs. A more insincere statement can hardly be imagined, and surely Dr. Condon, master in the politico-scientific world, would be the first to recognize it as such. For one could easily imagine the plight of a government funding agency, always hard pressed for funds, were they to support such research in the face of Condon's crushing summary of the situation. There would quickly be scathing howls of complaint and letters to Congress from rejected applicants for support in established scientific fields, asking why their proposals were turned down while "this UFO nonsense" was being supported.[2]

The rest of the lengthy report defies succinct description. It is a loose compilation of partially related subjects, each by a different author, but some sections do deal with direct investigations of selected UFO cases. It is these that, read carefully, give the lie to the Condon summary. Thus buried in the report, one finds many provocative statements, as, for example: "In conclusion, although conventional or natural explanations certainly cannot be ruled out, the probability of such seems low in this case, and the probability that at least one genuine UFO was involved appears to be fairly high." And in another instance: "This must remain as one of the most puzzling radar cases on record, and no conclusion is possible at this time." Again: "It does appear that this sighting defies explanation by conventional means." Another: "The three unexplained sightings which have been gleaned from a great mass of reports are a challenge to the analyst." And to cap the explanation gap syndrome, which weaves through the report, is this revealing remark: "This unusual sighting should therefore be assigned to the category of some almost certainly natural phenomenon which is so rare that it apparently has never been reported before or since." (How did this rare event get into the only ninety out of a potential 25,000 that could have been examined. How many similar rare events lurk in the remaining 24,910 reports?)

The thesis of the present chapter is simply that (a) the subject matter for study by the Condon group was incorrectly defined, and (b) the committee studied the wrong problem.

The UFO was defined by Condon as merely something that puzzled a given observer. The "Condon UFO" was not required to undergo a screening process before being admitted for study as a UFO: a report that *remained* unexplained after severe screening by technically aware persons. The committee thus really addressed itself to the problem of finding a natural explanation to fit the report. It is my contention that this should have been done in the original screening process. The fact that more than 25 percent of the cases studied were not assignable to natural causes simply means that only 25 percent of the cases studied were eligible for study as UFOs.

It was these cases (and many others that the Air Force had labeled Unidentified) and only these that should have been given continued study. The history of science has shown that it is the things that *don't* fit, the apparent exceptions to the rule, that signal potential breakthroughs in our concept of the world about us. And it was these cases that should have been studied from many angles. The committee chose to consider only the problem of whether UFO reports (and far many more nonUFO reports) supported the hypothesis that the earth was being *visited* by extraterrestrial intelligences [ETI]. UFO = ETI was the defining equation. It did not try to establish whether UFOs really constituted a problem for the scientist, whether physical or social. The question of whether puzzling reports of UFOs throughout the world might constitute "genuinely new empirical observations" was not considered. Thus the committee really studied the problem of misperceptions and their misinterpretation as evidence of extraterrestrial visitation. Perhaps this is a problem for sociologists and psychologists, who might well be interested to know that many thousands of people cannot identify Venus, a meteor, or an aircraft landing light and interpret the misidentification as visitors from outer space.

The problem was—and remains—whether the phenomenon of UFO reports from more than one hundred countries represents something genuinely new to science, quite apart from any preconceived theory (such as ETI) to account for the reports.

No critique of the Condon Report can avoid mention of the choice of data for study. By concentrating largely on current cases (forty of the ninety cases studied were in the year 1967) and also on relatively few cases out of the thousands available to them, they could not pay attention to the worldwide patterns of sightings during the previous twenty years. There were more than 12,000 Air Force reports available to the committee as well as the many thousands in the NICAP and APRO files (the latter were not made available to the committee largely because of the exceedingly poor psychological approach to APRO made by the committee). Dr. Saunders has remarked that in his statistical studies of UFO reports (not included in the Condon Report because Saunders was fired from the committee) he found little overlap in the Air Force and NICAP files except in the case of highly publicized cases. From my knowledge of APRO files I believe the same would apply there.

Thus, even though the emphasis on the use of current cases could be defended, the validity of this procedure rests on the assumption that these (and the some fifty others) were representative of the 25,000 cases in various files. For instance, only a few of the cases used in building the prototypes of categories in this book were studied, and of these exactly none was explained. (I recognize that I could successfully rest my case at this point.)

The Condon Report settled nothing. However, carefully read, the report constitutes about as good an argument for the study of the UFO phenomenon as could have been made in a short time and by a group of specialists in their individual disciplines having no prior knowledge of the subject.

Some knowledge of how the Condon committee came to be is important to the understanding of its actions. In a very real sense one can say that the Condon committee had its origin in "swamp gas." When in 1966 I suggested swamp gas as a possibility for the origin of that portion of the numerous Michigan sightings at Dexter and Hillsdale, in which *faint* lights over swampy areas were observed (the explanation was never intended to cover the entire spectrum of stories generated in that general area at that time), swamp gas became a household word and a standard humorous synonym for UFOs. UFOs, swamp gas, and I were lampooned in the press and were the subjects of many a delightful cartoon (of which I have quite a collection). Had a UFO been reported at that time from the Sahara Desert, I think it would have been attributed to swamp gas.

The good citizens of Michigan did not relish the raillery that developed, and a bipartisan Congressional pair, Weston Vivian, Democratic Congressman from Ann Arbor and House Republican minority leader Gerald Ford, called for a Congressional hearing into the matter.[3] A few quotations from the hearing are relevant of interest here:

THE CHAIRMAN (L. MENDEL RIVERS): Dr. Hynek, is there anything you would like to say to us?

HYNEK: Mr. Chairman, the press has treated me rather unkindly.

CHAIRMAN: You ought to be chairman of this committee.

HYNEK: The press has described me as the "puppet of the Air Force" and has stated that I say only what the Air Force tells me to say. I would like to . . . read to the committee a statement . . . which has certainly not been dictated by the Air Force.

CHAIRMAN: At this point, I want you to turn the loudspeaker up up . . .

HYNEK: . . . the kind of activity that the press has reported in Michigan is not unusal. It happened only that the Dexter and Hillsdale incidents, although of little scientific significance, have attracted national interest. Now, similar incidents, and some considerably more intriguing, have been occurring for many years. . . . Despite the seeming inanity of the subject, I felt that I would be derelict in my scientific responsibility to the Air Force if I did not point out that the whole UFO phenomenon might have aspects to make it worthy of scientific attention. . . . I am happy that my appearance before this committee affords me a chance to reiterate my recommendations. Specifically, it is my opinion that the body of data accumulated since 1948 deserves close scrutiny by a civilian panel of physical and social scientists and that this panel should be asked to examine the UFO problem critically for the express purpose of determining *whether a major problem really exists.*

CHAIRMAN: You say you can't write these reports off, you can't ridicule those who have made them. They are highly responsible people, in various walks of life, that have reported them [it is interesting that many of these words were original with the chairman. He had told us a short time earlier that his wife was favorably interested in UFOs] . . . Now, are you saying to us this morning that there should be a panel set up of scientists authorized by the Air Force before whom these things may be brought, and from whom a report should come?

HYNEK: Yes, sir, I am saying that. That would be the gist of my statement. However, I have been scooped by Secretary Brown, who has mentioned that the Scientific Advisory Board has recommended the same thing.

Just weeks before a special committee of the Scientific Advisory Board, under the chairmanship of Dr. Brian O'Brien, had recommended among other things: "Contracts [must] be negotiated with a few selected universities to provide scientific teams to investigate promptly and in depth certain selected sightings of UFOs. . . . The universities should be chosen to provide good geographical distribution. . . ."

The O'Brien committee had been called into being by a letter from Major General E. B. LeBailly, USAF Director of Information, which stated in part:

> Accordingly, it is requested that a working scientific panel . . . be organized to review Project Blue Book and to advise the Air Force as to any improvement that should be made. . . .
>
> Doctor J. Allen Hynek, who is chairman of the Dearborn Observatory at Northwestern University, is the scientific consultant to Project Blue Book. He has indicated a willingness to work with such a panel in order to place this problem in its proper perspective. Dr. Hynek has discussed this problem with Dr. Winston Markey, the former Air Force chief scientist.

A short while before that I had received a letter from Lieutenant Colonel J. F. Spaulding of the Air Force Office of Information concerning this problem, to which I had replied, in part:

> I have thought a great deal about your letter of August 13 [1965], in which you raised the question of exploring with the National Academy of Sciences the possibility of their looking into the UFO problem. . . . In the first place, the idea of having a civilian organization assist the Air Force in the UFO problem, either by working with them or by taking it over entirely, is not a new one. It has come up several times in the past eighteen years. In 1952, the Battelle Memorial Institute, in Columbus, was given the task of making a statistical study of the UFO reports up to that time. I was not at that time called in as a consultant, but during the time that Colonel Friend was in charge of Blue Book, we devised a panel of scientists, culled from Wright Field, which met regularly to assist in the evaluations. If I remember correctly . . . [we included] a psychologist and even a chaplain! But because this was an in-house effort, with no backing from the top, the panel was shortlived.
>
> Still later, SAFOI or its equivalent considered making overtures to NASA and to NSF for similar assistance, but after a few meetings . . . nothing came of it. With the exception of one further attempt, to interest the Brookings Institution into possibly looking

> into the matter, the problem has always remained an Air Force concern, and, I would say, is likely to remain so. . . .
>
> It is now, therefore, my considered opinion . . . that a civilian panel of scientists . . . be asked to examine the UFO problem critically for the express purpose of determining whether a major problem really exists. . . . The panel should be a working panel . . . whose members are willing to do a fair amount of "homework" between meetings.
>
> I would, of course, be willing to assist such a panel in whatever way I might and would even be willing to take a short leave of absence from my university if it would help place this problem in its proper perspective.

Copies of his letter went to Dr. Winston Markey, Chief Scientist, USAF, and to Dr. Harold Brown, Secretary of the Air Force. The use of several similar if not indentical phrases in Colonel LeBailly's letter a few weeks later to the Scientific Advisory Board indicates that my letter was at least read in the proper places.

So, it came to pass (after several abortive attempts to place the UFO study at an Ivy League university) that the University of Colorado accepted the challenge on October 6, 1966. I was not to be a member of the study group, possibly on the grounds that the committee should be composed only of those who knew nothing of the subject and hence could "take a fresh look" at it. This seemingly laudable criterion had its own built-in dangers and was, in a sense, like asking a group of culinary novices to take a fresh look at *haute cuisine* and open up a threestar restaurant.

Nonetheless, I understood the rationale, and, originally, I was pleased with the prospects. I remember a most pleasant meeting at the home of Dr. Franklin Roach, one of the committee members and my long-time astronomical colleague, at which Dr. Condon and several other committee members were present. There seemed to be a sense of adventure such as one gets at the start of a long journey. I recognized even that evening, however, Dr. Condon's basically negative attitude (and that of Mrs. Condon, which was particularly strong) but felt that this was only the natural skepticism on the part of a scientist who had not yet examined the data. I had no inkling then of how sparse and how poorly selected those data were to be.

Before long, as the committee began its work, I began to hear disturbing stories, first from one, then another of my friends associated with the Colorado project. There seemed to be real difficulty in defining the problem: the three psychologists differed sharply as to what the committee should study. One of them insisted that people were "just seeing things," and he would not entertain for one moment that an actual physical phenomenon worthy of study could possibly be involved. In his questionnaire he devoted one page to the elements of the sighting and twenty to the observer's psychological reactions.

Another psychologist became enamored of the idea that the whole problem of UFOs was the ETI hypothesis. Still another, concurring with this theory, held forth the impossibility of ever distinguishing ETI, if it existed, from a catch-all miscellaneous category.

One of the physical scientists proposed the use of a stereo camera over one lens of which would be placed a diffraction grating so that the spectrum of UFO lights could be determined. He did not know that the same idea had been proposed and put into limited use in 1954 by Dr. Joseph Kaplan, one of the organizers of the International Geophysical Year. I had demonstrated shortly after that by actual tests that the device was inadequate except for extremely bright lights. I still have one Videon stereo camera issued at that time to air bases, forlorn souvenir of the old days of UFOs. So much for a fresh look at the subject.

Dr. Saunders espoused the ETI hypothesis as the central solution for which to be tested. It certainly was the most spectacular thing to go for, even though there was no real evidence that it constituted the basic problem. However, in defining this to be the central question about UFOs, the committee was going along with popular opinion. In the public mind UFO is virtually synonymous with space visitors, generally regarded as "little green men."

However, Saunders' approach to the problem, once defined and adopted, was excellent: in the briefing of the Condon committee, which Jacques Vallee and I had been invited to make in the first weeks of its existence, we had both strongly urged that all available data, suitably weighted, especially those in the Air Force files, be put into machine readable form so that electronic computers could be used in data analysis. Saunders began to computerize the data available, and at the time of his firing from the committee he already had several thousand cases on magnetic tape. (At this writing he has some 30,000 cases on tape, available for sophisticated analysis.) Because Saunders fell out with Condon, none of his statistical work was included in the report, which "satisfied" its readers by an analysis of some ninety cases, many current, and in many cases did not satisfy the definition of UFO adopted here. Yet Vallee and I had virtually pleaded with the committee to seek patterns in thousands of reports, that it was essential to obtaining an overall perspective of the UFO situation. Only in that way could they see what sorts of things were being reported and by whom. Instead, the Condon committee dug in on a distressingly sparse sample of cases without knowing where these specific cases fitted in the overall picture. Were they truly representative of the really puzzling reports?

Groping for a methodology was an absorbing pastime for the committee. Although the report is entitled Scientific Study of Unidentified Flying Objects, was it indeed scientific? Or, indeed, could it be? Can the standard methods of science, so successful in areas in which the experiences are repeatable under controlled laboratory conditions, be applied to the UFO phenomenon?

It has been said that it is not the subject matter but the methodology employed that determines whether a study is scientific. By and large this can be accepted, but is it true in this particular area—the UFO? Granted that the best UFO reports are incredible tales told by credible people, how can one study them except to analyze, classify, and describe in more precise and orderly terms what was already contained in the report. What new insights, new evidence can be introduced except further details substantiating an already unbelievable tale?

In accepted scientific procedure one generally has some hypothesis to test. "If—then" is the epitome of the scientific method. If this is so, then that will follow, and the hypothesis means nothing unless the "then" is testable, can be shown to be true or false. Particularly important is the latter, that a hypothesis can unequivocally be shown to be false. Is there some crucial experiment that can be performed or some observation made that will prove the hypothesis false? If not, how can one distinguish between one hypothesis and another?

The Condon committee chose to test the ETI hypothesis, that UFOs were solid evidence of extraterrestrial visitation. How can this be shown to be false? Elaborate observational networks might be set up, and no UFOs show. Is that negative proof? No. One can always say that the intelligently guided UFOs knew they were expected and therefore avoided the trap.

True, the Condon committee found nearly a third of their cases without even partially adequate explanation. The "experts" were stumped. What could that possibly say about the ETI hypothesis? Nothing. There could be many explanations, depending on how bold one wishes to make their initial assumptions. For example, if one wishes to postulate worlds other than the physical (astral or etheric), one can easily satisfy and explain virtually all the reported antics of the UFO. But how do you establish that the hypothesis is true? Unless you have an operational method of doing so, it is not science. Hypotheses remain hypotheses, and we are left with "the two-and-seventy warring sects."

Even if the Condon committee had had orders of magnitude more data, they tackled a hopeless task. The only hypothesis the committee could have productively tested was: *there exists a phenomenon, described by the content of UFO reports, which presently is not physically explainable.* That hypothesis is capable of being proved untrue by the simple expedient of explaining, by present physical principles, the thirty or so cases they were unable to treat satisfactorily, and, of course, the many hundreds of others not treated by the Condon group. Even so, one can always bring in new cases, saying, "Here, you haven't explained this one," but reason dictates that if a predetermined number, *n*, of cases (submitted by a panel of persons well versed in the subject, who have subjected the cases to careful screening to virtually exclude birds, balloons, aircraft, meteors, planets, etc.) can be explained, the issue is closed unless new observational and dissimilar data are presented. Conversely, if the test cases cannot be so explained, quite obviously there exists a phenomenon, almost by definition, that is not explained by present physical principles.

A careful reading of the report establishes that the committee went a long way, inadvertently, to establish the nonexplainability hypothesis while floundering with the ETI hypothesis. By their own admission, their "experts" were indeed stumped by many of the relatively few cases examined, and there was no indication that they would have been successful had they tried their hand at other cases for which more knowledgeable investigators had also failed to provide a physical explanation.

In my opinion, it is most unfortunate that events transpired that led to the ultimate dismissal of Dr. Saunders. The course of the committee would surely have been different had he remained and had his counsel been accepted. Even though he had focused on the ETI hypothesis for testing, he would soon have recognized that while the hypothesis was impossible to establish or disprove, the "unexplainability by present physical principles" hypothesis was obvious.

Dr. Condon evidently was aware of Dr. Saunders's worth to the project, for before the dismissal, when applying for additional funds ($259,146 in addition to the original $313,000), he had written in the official proposal:

> Dr. Saunders has many duties. He has been directing the acquisition, cataloging, and organization of the sightings file. . . . Saunders has been responsible, with the assistance of the other Study Team members, for the development of the interview and sighting report forms. . . . Dr. Saunders is also responsible for the statistical analysis of UFO data. As this is written he has completed the punching of some 12,000 sighting reports for machine study. It is intended that additional data will be punched for many of the sightings when techniques have been developed for coding a number of parameters that are difficult to handle statistically. . . . In order to promote the articulation of these interrelated functions, Saunders has been given the responsibility for all of them: the sighting record keeping, the statistical analysis, and the formulation of the questions asked on the interview and sighting report forms. There is another related function: decision-making leading to the signal to send investigating teams into the field to study reports of sightings. Saunders is also responsible for that.

Shortly after that letter was written, Condon fired Saunders for incompetence. Seems quite a lot of responsibility to have assigned to an incompetent.

The events leading up to the dismissal of Saunders and another committee member, Dr. Levine, and the resignation shortly thereafter of Mary Louise Armstrong, Condon's administrative assistant, are treated in Saunders' book, to which I have already referred.

Instead of extracting the essence out of many thousands of cases, which Saunders was well on his way to doing, the report contains only eighty-seven investigated cases, plus three unexplained sightings from astronauts (the investigator of the sightings stated laconically that they ". . . are a challenge to the analyst").

The Condon Report was released on January 9, 1969, simultaneously with the seal of approval of the National Academy of Sciences. The latter release concluded:

> In our opinion the scope of the study was adequate to its purpose: a scientific study of the UFO phenomena.
>
> We think the methodology and approach were well chosen, in accordance with accepted standards of scientific investigation,

These statements imply that the scientific method is indeed applicable to the UFO problem, a point I have already questioned. Since the hypothesis the committee was testing (ETI) is nonfalsifiable, that is, would be impossible to prove its negative, a possibility insisted upon by the scientific method, the method is not applicable unless the problem is properly redefined.

However, giving the academy the benefit of the doubt, the methodology of the Condon committee is easily faulted in other areas. Let us speak here of the methodology without confusing it with the subject. My criticisms of the methodology of the Condon committee would be the same had the subject not been UFOs but the life cycle of the gray whale (if the study had been directed to the testing of just one theory, perhaps that the gray whales were products of special creation) or of the causes of cancer (had the study been limited to the theory that cancer is caused by improper diet).

The academy would agree that the scientific method implies that *the specific problem to be studied must be defined and must be relevant to the larger field that contains it.*

On page 9 of the Condon Report, the UFO is defined: "An unidentified flying object is here defined as the stimulus for a report made by one or more individuals of something seen in the sky (or an object thought to be capable of flight but seen when landed on earth) which *the observer* [italics mine] could not identify as having an ordinary natural origin, and which seemed to him sufficiently puzzling that he undertook to make a report of it. . . ." And the problem is defined: "The problem then becomes that of learning to recognize the various kinds of stimuli that give rise to UFO reports."

Scientific method! What sort of a scientific investigation is it that assumes the answer before starting. The assumption here is clearly that UFOs are all misperceptions of natural things and that the entire work of the committee was to learn and memorize the varieties of natural stimuli for UFO reports so that all one needed to say was, "That must have been Venus; that one must have been an airplane landing light." There was no room in the Condon concept of the problem for even the possibility of the statement: "That was probably a UFO."

Thus this principle of the scientific method was violated: the problem was ill-defined and did not relate to the larger field, the truly puzzling reports for which the generating stimulus was not obvious. Further, it cannot be left to the observer, who represents in general a cross section of the entire populace, to define the problem, to attach the label *UFO*. That can only be done by those capable of critical screening out of precisely those reports for which the Condon definition holds—those that are generated by natural stimuli.

In the larger sense, the problem certainly is to find the stimulus for the UFO report. But to assume at the start that the generating stimulus must of necessity be only of one class—misperceptions—is indeed a violation of the scientific method. It would seem that the committee defined the problem as one of finding natural stimuli, yet it chose to test the hypothesis of extraterrestrial intelligence. Venus is not an extraterrestrial intelligence; a meteor is extraterrestrial but it certainly is not intelligent and so on.

Another prime principle of the scientific method with which the academy would agree is: *the data chosen for the study should be relevant to the problem.*

The question of relevance can be referred either to the problem the committee investigated (extraterrestrial visitation) or to the one they did not investigate: do UFO reports, properly screened, signal the advent of empirically new observational data?

In either case, the majority of reports actually used was not relevant; an experienced investigator would have screened out the obvious misperceptions that were puzzling to one or two persons but would not have fooled an expert. In fact, fourteen of the eighty-seven cases had previously been evaluated by Blue Book as misperceptions, yet Condon chose to let the relatively few cases examined be diluted by trivial cases. It would have been better if those fourteen had been replaced by fourteen cases of the several hundred Blue Book had classed as Unidentified. It was in such cases that the solution to the problem would lie, if at all.

Only ten Close Encounter cases, certainly the most interesting of all UFO reports, were examined by Condon's group. Of these the committee was unable to explain six to any degree, two were considered inconclusive, one psychological, and one was most definitely Venus! The latter case should be read by all UFO investigators. It is a fantastic example of how persuasive the planet Venus can be as a nonscreened UFO. Police officers in eleven counties were "taken in" by this planet. It is a case of particular value to psychologists and, one is tempted to say, to those responsible for the hiring of policemen.

It has long been my experience that no case involving the appearance of a "UFO" more or less on schedule night after night should be taken seriously. It is almost certain to turn out to be a scheduled aircraft or a planet, especially if one is told it didn't appear on cloudy nights. Such cases, however, are so easy to filter out that they provide interesting comic relief to an otherwise perplexing problem.

The academy, I believe, would agree it is good scientific methodology *to avoid bias, prejudice, and ridicule in approaching a problem.* The answer to a research problem should never be anticipated to the degree that it strongly influences the approach to the problem. In my contacts with the dozen or so committee members and associates in the Colorado study group with whom I had the pleasure of speaking, I found no pronounced bias in their approach.

There were differences of viewpoint, of course, but no corrosive emotionally charged bias. If one judges the director of the project, however, solely by his actions and spoken and written word, this does not appear to have been the case. Always outspoken, he did not hesitate to reveal his inner attitudes in the talks he gave from time to time in various parts of the country. One of the first of these, delivered when the project was less than three months old, was at Corning, New York (reported in the January 26 issue of the Elmira, New York *Star-Gazette*). Condon reportedly said, "It is my inclination right now to recommend that the government get out of this business. My attitude right now is that there's nothing to it . . . but I'm not supposed to reach a conclusion for another year. Maybe it [the UFO problem] would be a worthwhile study for those groups interested in meteorological phenomena."

Every man is entitled to his own opinion, but a scientist carries an additional social responsibility by virtue of his position and profession. His words, especially those idly spoken, can carry undue weight in the public press. Here we have Condon saying the project is not worthwhile (a few months later he asked for $259,146 extra to continue the work) and then displaying his conviction that UFOs must, of necessity, be natural (meteorological) phenomena, with the implication that there was no point in looking further.

Much later in the year Condon spoke at the National Bureau of Standards in Washington. "According to the reports that came from members of that audience, and by his own later admission, Condon concentrated almost the entire talk on three of the crackpot cases with which he had been involved."

I think the National Academy would also agree in its appraisal of the application of the scientific method that *no scientist should willfully allow ridicule to be an accepted part of his scientific method.* When, however, a subject seems to be beyond the pale of science (and history is full of examples), raillery and banter at the expense of the other fellow does not bother the scientist's conscience. Thus, Dr. Menzel's written reply on a serious questionnaire which asked, "what should be done about UFO reports that can't be explained," was, "Throw them in the wastebasket!"

Dr. Condon apparently felt UFOs beyond the pale of science (even though his report is entitled Scientific Study of Unidentified Flying

Saucers), for he too resorted to banter and jokes at the expense of the other fellow. Saunders points out[5] Condon's preoccupation with the kooky aspects of the UFO problem and his seemingly callous ridicule of associated persons (even though it has been well established that the "kooks" *do not* generate articulate and coherent UFO reports). Saunders remarks, "Worst of all, his treatment of the persons in these cases offended me as a psychologist. They may have needed help, but they did not need to be laughed at. It was as if Condon had lost all sense of perspective and was sacrificing these unfortunates to relieve his own frustration. It seemed that as soon as Condon had gotten as many laughs as he could from one case, he would immediately top it with another." In one instance, Saunders relates in his book, Condon phoned the governor of Utah to apprise him of the predicted landing (by a person "in contact" with extraterrestrials) of a rectangular-shaped UFO on the Salt Flats at Bonneville.

In another instance he passed information to Washington with a straight face about an offer (for three billion dollars) made to him by "an agent of the Third Universe" to construct a spaceport so that ships from this universe could land on our world.

I confess that the temptation to get a laugh out of the antics of the lunatic fringe is great. I have used to good advantage, as comic relief in an otherwise serious lecture, a photograph illustrating a story that appeared in a popular magazine entitled, "A Flying Saucer Saved My Virginity," a cartoon of the Three Wise Men gazing up at the star, one saying to the others, "Swamp Gas," and other bits of UFO froth.

However, as long ago as 1953[6] I wrote:

> Ridicule is not a part of the scientific method, and the public should not be taught that it is. . . . The steady flow of reports, often made in concert by *reliable* observers, raises questions of scientific obligation and responsibility. Is there, when the welter of varied reports are shorn of, in the words of Poo Bah, all "corroborative detail to lend artistic verisimilitude to an otherwise bald and unconvincing narrative," any residue that is worthy of scientific attention? Or, if there isn't, does not an obligation still exist to say so to the public—not in words of open ridicule but seriously, to keep faith with the trust the public places in science and scientists?

The relevance of science in daily life has in our times been seriously questioned. Supercilious attitudes, pontifical *ex cathedra* statements, and demands that authority be worshipped just because a scientist said so—these things do not help. The public, from whom the support of all scientific endeavor ultimately must come, should be given the chance to see science as an adventure pursued in humility of spirit, with dignity and respect, and for the benefit of all. It should be emphasized that in science one never knows where inquiry will lead—("if we know the answers in advance, it isn't research")—that a primary aim of science is to satisfy human curiosity, to probe the unknown, and to open new paths for intellectual adventure. This is in line with what science has always stood for, even though scientists, being quite human, have often inadvertently given quite the opposite impression.

I believe the National Academy would agree that it is in keeping with the scientific method *that the director of a scientific project should understand the problem.* Unimpeachable evidence that Condon did not understand the nature and scope of the problem is given in the examples of "UFOs" he used to support his summary of the report. I quote here with permission from a paper by W. T. Powers[7]:

> Condon uses as illustrations exclusively the cases which are silly, easy to explain, or poorly reported. There is not one word about the fact that his colleagues present, in the same volume, cases which resisted the most meticulous attempts at explanation and which were not silly, poorly reported, or easy to explain.
>
> Condon concludes his section on "'radar sightings" of UFOs without mentioning the cases for which his own staff could find no substantiation for "anomalous propagation" by saying, "In view of the importance of radar to the safe operation of all aircraft, it is essential that further research be done. . . . However, it is felt that this can be done by a direct attack on the problem . . . rather than by investigation of UFO cases."
>
> In short, Condon does not admit that radar reports of UFOs can tell us anything about UFOs—all such reports can do apparently is to reveal anomalous propagation. The possibility of radar sighting of a UFO by a properly functioning radar set under normal atmospheric conditions is not mentioned at all, even as a possibility. Yet

there are *three* such carefully studied cases among the reports later in the text.

It is instructive to examine the individual cases which Condon chose to illustrate the points made in his summary. To avoid forcing conclusions, I will list and comment briefly on *all* examples used by Condon so as not to commit the sin that is being criticized. [Space does not permit quoting them all here.] The first example is a lights-at-night case (hot air balloon), so is the second (Saturn). Neither would have survived as a UFO the first pass by an experienced investigator. . . . Another example concerns a man whose ancestors came from another galaxy . . . another concerns the planet Clarion, a fantasy indulged in by certain fringe groups of half UFO buffs and half psychic believers. This merits a whole page!

Then we come to photographs. A page is devoted to a photo determined to be either an error or a hoax. Of a case which his photographic expert took quite seriously, Condon says only "the UFO images turned out to be too fuzzy to allow worthwhile photogrammetric analysis." Yet later in the text: "This is one of the few UFO reports in which all factors investigated, geometric, psychological, and physical, appear to be consistent with the assertion that an extraordinary flying object, silvery, metallic, disc-shaped, tens of meters in diameter, and evidently artificial flew within sight of two witnesses." Description of the photogrammetric analysis occupies pages 399-407 of the report. Condon concludes this section of his summary with a brief mention of a photo of a "bear track" and a lens flare [on the cover of the Bantam edition of the Condon Report is found a pretty color picture of a lens flare, a photograph that would never have been taken seriously by an experienced investigator]. No mention is made by Condon of the exhaustive analysis made of the Great Falls, Montana, movies made by Dr. R. M. L. Baker, an analysis which had been submitted to the Condon committee.

Automobile malfunctions are next. Condon states incorrectly that only one case came to the attention of his group. . . . He does not mention the other case [case 12] of reported automobile malfunction [nor indeed any of the hundred or more such cases available to him had he bothered to survey the literature. It is standard in scientific methodology to conduct a literature survey before an investigation

> is undertaken in order that the investigators will become conversant with what has gone before and thus reduce the chances of needless duplication] in which the witness was described as a "competent, practical personality, trained and accustomed to keeping her presence of mind in unexpected situations."
>
> Condon does mention the astronaut's visual sightings but says only that "nothing was seen that could be construed as a flying saucer" or manned vehicle from outer space, even though he admits that Dr. Roach, the astronomer who made a thorough investigation of the astronaut's sightings, had termed these sightings a "challenge to the analyst." Condon was not interested in taking up the challenge even though Dr. Roach had stated, "Especially puzzling is the first one on the list, the daytime sighting of an object showing details such as arms protruding from a body having noticeable angular extension. If the NORAD listing of objects near . . . at the time of the sighting is complete, as it presumably is, we shall have to find a rational explanation or, alternatively, keep it on our list of unidentifieds." Apparently Condon was not interested in either alternative.

Powers includes many more examples and remarks:

> These latter examples show the strong selection factor in Condon's summary—none of them concerns the kind of UFO report that would keep the attention of an experienced investigator. It is evident that Condon systematically avoided bringing up as illustrations the most puzzling cases in his report and that he systematically misrepresented those few puzzling cases which he did mention, in the direction of playing down or ignoring what was unexplained and playing up possible explanations even when the detailed analysis all but rules them out.
>
> If Condon really wanted to take a physical science approach, why did he not investigate whether or not an unknown physical phenomenon was responsible for some well-chosen class of UFO reports? Why did he waste his time and our money chasing after lights-in-the-sky reports and reports of Venus, and, especially, why did he set up that straw-man ETI hypothesis? If we don't even know whether a phenomenon exists, how can we possibly test any wild guess about its cause?
>
> Condon was responding not to fellow scientists who have indicated a possible value in UFO research but to the claims of contactees,

> to the excesses of ignorant believers, to the naive questions of the uninformed. He did not choose to deal with the problem beginning at the level to which others, just as competent as he, had brought studies of UFOs. Instead he attacked the ideas of those who are easy to attack. . . . Condon's report and especially his own comments . . . are an affront to his fellow scientists. It is perfectly clear, however, that the summary he wrote is in fact biased, and that the National Academy of Science has been thoroughly misled.

Those words of Mr. Powers are strong indeed, and it is clear why the establishment editors of *Science* refused to publish them. It should be remembered, however, that Mr. Powers was not discussing UFOs; he was discussing scientific methodology and asking whether the methods of science were used in the Condon Report, as the National Academy had vouched was indeed the case. Perhaps we might ask whether the committee appointed by the academy to review the report truly did its homework. We could more easily excuse them if they had not. Condon had many other duties and had never expected to devote full time to UFOs. He appointed Mr. Robert Low to be his project administrator, and as time went on and Condon became more and more out of touch with the committee, Low became the actual pilot of the Colorado UFO ship. It is conjectural how the project might have turned out had a different scientific administrator been chosen. I remember my own dismay when, on the occasion of my visit to the committee, when the project was scarcely two weeks old, Low outlined on the blackboard for us the form the report would take, what the probable chapter headings would be, how much space should be devoted to each chapter, with an implied attitude that he had decided already what the substance and tone of the report would be.

It was Low who authored the famous memo, the instrument that led to the sacking of both Saunders and Levine.[8] The much quoted key message, written on August 9, 1966, nearly three months before the project formally began, was "The trick would be, I think, to describe the project so that, to the public, it would appear a totally objective study but, to the scientific community, would present the image of a group of nonbelievers trying their best to be objective but having an almost zero expectation of finding a saucer."

I believe Low has been unduly criticized for this memo. I can appreciate the dilemma Low faced. He wanted his university to get the contract (for whatever worldly reason) and to convince the university administration that they should take it. He was aware, as I certainly have been for years, that scientific opinion was such that even serious mention of the subject was the equivalent of scientific tar and feathers. He wanted to invoke a cloak of respectability. But the path he chose was unfortunate.

Saunders and Levine were fired for having called this memo to the attention of a few colleagues; news of its existence spread and led in time to the Fuller exposé in *Look* magazine.[9]

After Saunders and Levine were sacked, Condon's administrative assistant, Mary Louise Armstrong, who knew the innermost workings of the entire committee, had the problem of whether to continue to work for a man for whom she had lost a great measure of respect. Two weeks later she resigned from the project, stating her reasons in a thoughtful letter, which should be made a matter of record (see Appendix 3) because it provides an insight into the inner life of the committee to which future historians of science should have access.

I give an excerpt from her letter[10] as it relates to the points under discussion:

> Since it is apparent to the staff of the UFO project, as well as to you [Condon] that we are in a real dilemma over the disagreement and low morale within the study as a result of the last two weeks, I feel it is necessary to examine what, in my opinion, has been the primary cause of the problems that exist. . . .
>
> I think there is an almost unanimous "lack of confidence" in him [Low] as the project coordinator and in his exercise of the power of that position. . . . Bob's attitude from the beginning has been one of negativism. Bob showed little interest in keeping current on sightings, either by reading or talking with those who did. . . . Saunders carefully set aside reports on a check-out basis, so that everyone on the committee would have a chance to read them and certainly never encouraged the proposed discussions to actually take place. . . . To me, too much of his time has been spent in worrying about what kinds of "language" should be used in the

> final report so as to most cleverly avoid having anything to say definitive about the UFO problem. Very little time, on the other hand, has been spent in reviewing the data on which he might base his conclusions. . . .
>
> Why is it that Craig, Saunders, Levine, Wadsworth, Ahrens, and others have all arrived at such radically different conclusions from Bob's? It is not my impression that they came into the project with any particular bias concerning the UFO problem. I think that there is fairly good concensus among the team members that there is enough data in the UFO question to warrant further study. That is not to say, as no one of us would, that we are definitely being visited by vehicles from outer space. . . . A dialogue will have to occur eventually in which both sides of the question are debated within the group but to be putting these ideas down on paper in the form of conclusions and discussing them with people outside the project is presumptuous and wrong. . . . I am impressed by the fact that it seems as if he is trying hard to say as little as possible in the final report, but to say it in the most negative way possible. I quote Dave Saunders when I say that Bob's suggestion that we could use footnotes for any minority opinions evoked Dave's response, "What do we do—footnote the title?"

One does get the feeling that somehow the slate should be wiped clean and the job done over—properly.

CHAPTER THIRTEEN

The Case Before Us

La Commedia é finita!

—*CANIO,* PAGLIACCI

The comedy, indeed, should be finished, and serious work should begin. The problem of UFOs *can* be attacked productively, and a positive program can be set down. Let us first be entirely clear as to what the problem is by summarizing what the previous chapters have shown and what they have not attempted to prove or establish. I would hold that it *has been established that:*

1. There exists a phenomenon, described by the contents of UFO reports (as defined here), that is worthy of systematic, rigorous study. The extent of such a study must be determined by the degree to which the phenomenon is deemed to be a challenge to the human mind and to which it can be considered potentially productive in contributing to the enlightenment and progress of mankind.
2. Even allowing for the unfortunate and disorganized manner in which the data have become available for study, the body of data points to an aspect or domain of the natural world not yet explored by science.
3. For a directed, objective study of the phenomenon the available data require major organization, systematization, and the adoption of a uniform terminology for their description and evaluation. Such

organization and systematization must be applied in the gathering and processing of new data.

4. Investigations that have sought to disprove the above have failed to make a case. Blue Book and the Condon Report are the principal examples of such fruitless efforts.
5. The probative force of the four uncontestable statements above strongly suggests that new empirical observations exist that describe a *new fact*—the existence of UFOs (as defined here)—which needs to be brought within an acceptable framework of concepts and, if possible, explained. Further work of an unbiased character is clearly the next step.

It is likewise important to keep clearly in mind what the previous chapters have not attempted to establish, prove, or show. It has not been shown:

1. That the new fact implied in (5), above, requires a basic shift in our outlook on the natural world.
2. What a verifiable explanation of the UFO phenomenon is. An organized approach to the problem must be formulated. In outline, the following steps should be taken:
   (a) The problem must be rigorously defined, and extraneous aspects must be clarified and set apart from the main problem.
   (b) Feasible, tractable methods of attack must be outlined, with great care being taken to avoid involved, prohibitively costly, and open-ended paths (for example, the establishment of thousands of manned or automatic highly instrumented observing stations).

THE PROBLEM DEFINED

In order to define clearly the problem of UFOs the following steps must be taken:

1. To delineate with far greater precision than heretofore the parameters of the phenomenon. In other words, to characterize as definitely

as possible the *strangeness* of the phenomenon: what are the factors of strangeness that we find common to the various observational UFO categories? What, in short, is there to explain?

The problem is not, at present at least, to explain or to solve the UFO phenomenon. That, of course, is the ultimate problem, but indications are strong that we at present do not possess the knowledge to arrive at a final solution. But we do have the means at our disposal by study of highly selected and screened UFO reports to characterize explicitly what needs to be explained.

2. To determine with far greater precision than heretofore the ordinate of the S-P diagram, that is, the probability that the strangeness of the UFO phenomenon is as stated. In other words, from a study of UFO reporters from over the world, we must use Hume's "betting criterion" of belief that the reporters of the phenomenon were not totally and egregiously mistaken in what they reported.

From the evidence over the past years, an overview of which has been already given, I would have to say that I would bet a sizable amount that the screened reporters were reporting a new fact—UFOs.

This, then, is the problem: to construct with as much precision as possible an S-P diagram for those UFO reports that meet the screening criteria. How do we go about attacking it?

The mass of worldwide UFO reports can be handled in two ways: statistically, in the mass, or specifically, one by one.

With the numbers of UFO reports of high strangeness now counted in the thousands, a statistical approach can be very productive, and methods suggested by modem information theory are certainly applicable. Sophisticated methods of information retrieval, pattern recognition, and significance testing have served in a number of disciplines to extract that signal from the noise in situations that at first glance seemed hopeless.

A simpler yet more powerful method of demonstrating significance of patterns is to compare large groups of sightings of a particular category with a much larger population of the same category. An example is one such examination by Jacques Vallee. He compared statistically one hundred Close Encounter cases from Spain (he used his designation of Type 1, which includes all three of our categories of Close Encounters) with 1,176 cases in the same class reported from all parts of the world other than Spain. He compared the occurrence of cases in which the object was reported as seen on the ground, those seen near ground level (generally hovering or moving at about treetop level), those having occupants reported and the percentages of the latter reported on the ground, and near-ground cases:

	ON GROUND	NEAR GROUND	OCCUPANTS	NO OBJECT
1176 Non-Iberian cases	60%	35%	32%	5%
100 Iberian cases	53%	38%	25%	9%

Likewise, the distribution of the occupant cases among the ground and near-ground cases was almost identical. The "no object" cases refer to reports of humanoids whose craft was presumably hidden nearby.

A correlation such as this would be accorded high significance in recognized disciplines such as sociology or economics. It points strongly to the existence of invariants in sightings of a given category. Why otherwise should the significant Spanish sample (which included virtually all well--reported cases from Spain during the past decade) be so similar to the much larger worldwide (excluding Spain) group containing also virtually all well-reported cases in roughly that same period of time?

Any serious study of the UFO problem would of necessity include many such correlation and pattern studies. Studies by categories of sightings—intra- and inter-category correlation studies—to establish geographical, seasonal distributions (how are the various categories related in these respects?) and studies of the kinematics exhibited by the UFOs within each category (do Daylight Discs and Nocturnal Lights have the same proportion of rapid takeoffs, hoverings, and sharp turns?) must be made.

Within the Nocturnal Lights category, to cite just one of many possible approaches, are the reported color changes correlated with the manner of motion of the UFO? When rapid acceleration occurs, what is the predominant color change reported worldwide, and how does this differ, if at all, in reports from widely separated portions of the globe?

A serious scientific group engaged in such studies would, given access to the data in machine-readable form, soon demonstrate beyond any reasonable doubt whether there was anything substantive in the UFO problem. The Vallee correlation, above, if repeated in an appropriate manner in seeking for patterns among worldwide Radar-Visual and Daylight Disc cases, country by country, would compel recognition (if the correlation were positive) that the UFO phenomenon represented new empirical observations that (by definition of new empirical observations) are not encompassed by our present scientific framework.

It may well be asked why all this has not been done before. The subject has actively concerned us for more than a score of years. A moment's consideration, however, will show what an impossible accomplishment this would have been. Most recently, the Condon group spent a half million dollars ostensibly to study the subject scientifically, but the members did not even consider this approach. How then could private groups without funds, without data in usable form, and usually without scientific training essay such a task? Blue Book did not even remotely consider this approach despite the strong advice of their scientific consultant. Recall, too, that the many thousands of Blue Book cases were arranged in folders only chronologically, with no semblance of even the most elementary cross-indexing.

As was true of many other fields of study in their infancy, scientific respectability is won slowly, with comprehensive study possible only after the subject is accorded some measure of acceptance. But even if UFO reports were to cease as of this moment and no reports of acceptable criteria were to be submitted henceforth, it is my opinion that the data that now lie scattered about, if properly processed, could establish the substantive nature of the UFO phenomenon beyond reasonable doubt.

However, UFO reports have by no means ceased at the time of writing, although they receive very little attention in the press, particularly in the urban press. It is thus difficult to assess the level of UFO activity. Reports, especially from scientifically and technically trained people, are accepted for scientific record purposes by me and my colleagues in Evanston, Illinois, with the understanding that they will be for scientific use only. Private UFO investigation groups in many countries continue to receive reports, synopses of which are published in the literature.

The *second* potentially productive approach to the UFO problem is the examination, in depth, of individual multiple-witness cases, particularly those of recent origin. Concentration here on Close Encounter cases clearly promises the most return, especially Close Encounters of the Second Kind, in which the reported presence of physical evidence can yield quantitative physical data.

The individual case approach requires persons trained in interrogation who also have an intimate knowledge of the various manifestations of the UFO phenomenon and are able to recognize the characteristics of reports generated by common misperception. It is imperative that they be well acquainted with both psychology and basic physics.

If even a handful of such crack investigators were available and had immediate reaction capability so that within a day or two (preferably within hours) they could be on the spot of the reported UFO occurrence, they could, with the original reporters, reconstruct the circumstances of the reported event at the exact location of the event, perhaps under closely similar circumstances, and thus could obtain at least semi-quantitative data.

A skilled interrogator can extract valuable data from a case that is months—or even years—old. Experience has shown that the UFO event remains etched into the memories of the reporters and that, especially if the experience has been somewhat traumatic, usable and valid details can still be obtained long after the event. I have found that the greatest obstacle to the investigation of old cases is not fading memories of the witnesses but the frequent unavailability of the reporters. Because of increased mobility of persons and families the most recent address of a critical witness becomes difficult to obtain as time passes. On occasion, as McDonald demonstrated in the Lakenheath and Texas-Oklahoma cases,[2] years later the reporters were located as a result of great effort. In those particular cases the witnesses were found to be most cooperative.

A trained investigator is able to extract the maximum amount of information from the reporters, translating vague statements such as "it disappeared very rapidly" into "it accelerated within a second to an angular speed of 10 degrees per second and disappeared into the cloud cover in the west-north-west." Apparent sizes, colors, directions, state of the weather, direction of wind, position of the sun or moon or planets, and other such data can generally be ascertained if a *trained* investigator is on the scene as soon as possible. In this way what generally survives only as an anecdotal statement or an imprecise account of a frightening and unusual experience can be transformed into a far more precise account of the reported occurrence. The investigator should at all times attempt to locate independent witnesses to the reported event, even at the cost of considerable effort.

With dedicated study of carefully selected cases and communication of their results, perhaps at national or international meetings, investigators could soon answer the important question: is there a genuine UFO phenomenon that represents something truly new to science? Experienced UFO investigators will cry in anguish at this statement; they are so convinced that the UFO phenomenon constitutes genuinely new empirical data that they would regard the above as an elementary point that could easily be skipped. Nevertheless the fact must be proved much in the same way that it was finally demonstrated nearly two centuries ago that stones did actually "fall" from the sky.

The combination of a sophisticated statistical approach and the detailed studies of specific individual multiwitness cases would almost certainly establish whether or not UFOs are indeed new empirical observations heretofore unrecognized by science.

An approach such as this is essential for the resolution of today's confusing situation. Views range from those who consider the entire subject as nonsense (either from a priori considerations or in the belief that the Condon Report has been definitive) and hence refuse to devote even a moment to the examination of the data, to those who have examined the present data and are convinced on that basis that the UFO phenomenon represents a new field of science. This severe polarization of the issue can be dissipated only by concentrated study. How can such studies be pursued best?

We can start with the knowledge that the UFO phenomenon is global, that UFO reports persist in this and other countries despite the Condon

Report and the closing of Blue Book, and that many small groups of scientifically trained people, especially young scientists, are expressing interest in the subject and dissatisfaction with the manner in which it has been treated in the past. Some find it increasingly difficult to understand why the National Academy of Science fully endorsed the Condon Report and its methodology.

Long before the release of the Condon Report, the AIAA (American Institute of Aeronautics and Astronautics) asked two of its technical committees, the Committee on Atmospheric Environment and the Committee on Space and Atmospheric Physics, to establish a subcommittee devoted to the UFO problem. Dr. Joachim P. Kuettner, of the ESSA Research Laboratories in Boulder, Colorado, was asked to chair the committee. In the December, 1968, issue of the *Journal of Astronautics and Aeronautics*, one of the official journals of the AIAA, the committee stated (just before the Condon Report was released on January 8, 1969): "The committee has made a careful examination of the present state of the UFO issue and has concluded that the controversy cannot be resolved without further study in a quantitative scientific manner and that it deserves the attention of the engineering and scientific community."

In the same journal some two years later[3] the UFO subcommittee published an article entitled, "UFO: An Appraisal of the Problem." Very cautiously worded, it was nonetheless critical of the previous treatment of the UFO problem by the scientific community. Commenting on the Condon Report nearly two years after its publication, the committee stated:

> To understand the Condon Report, which is difficult to read, due in part to its organization, one must study the bulk of the report. It is not enough to read summaries, such as those by Sullivan and by Condon, or summaries of summaries on which the vast majority of readers and news media seems to rely. There are differences in the opinions and conclusions drawn by the authors of the various chapters, and there are differences between these and Condon's summary. Not all conclusions contained in the report itself are fully reflected in Condon's summary.

Later in the report of this committee we find:

> Condon's chapter, Summary of the Study, contains more than its title indicates; it discloses many of his personal conclusions. Making value judgments was no doubt one reason why Condon was asked to handle the project. One is happy to obtain the judgment of so experienced and respected a man; but one need not agree with it. The UFO subcommittee did not find a basis in the report for his prediction that nothing of scientific value will come of further study.

Still farther in the report we find:

> Taking all evidence which has come to the subcommittee's attention into account, we find it difficult to ignore the small residue of well-documented but unexplainable cases which forms the hard core of the UFO controversy.

The committee likewise concurred with my own feelings about the extraterrestrial hypothesis (ETH):

> We have already expressed our disenchantment with arguments about the probability of the extraterrestrial origin of UFOs since there is not sufficient scientific basis at this time to take a position one way or another . . . the UFO subcommittee feels that the ETH, tantalizing though it may be, should not be dragged into this consideration as it introduces an unassessable element of speculation; but the subcommittee also strongly feels that, from a scientific and engineering standpoint, it is unacceptable to simply ignore substantial numbers of unexplained observations and to close the book about them on the basis of premature conclusions.

The AIAA committee has suggested a proper first step in a new approach to the problem:

> The subcommittee sees the only promising approach as the continuing moderate-level effort with emphasis on improved data collection by objective means and on high quality scientific analysis.

The general confusion surrounding the subject and the lack of attention by scientists have effectively prevented proper data collection. Even after twenty years of sporadic, unsystematic data collecting there exists only a formidable collection of heterogeneous data, often consisting of little more than discursive, anecdotal accounts. The more than 12,000 Air Force cases are arranged only chronologically, with no attempt at cross-indexing, and the same is true of the files of many private investigators and organizations.

Thus the first step means starting almost from scratch: *data gathering* and *data processing*. This may seem to be a most pedestrian approach to a most exciting topic, but so far we have only an airy, unsubstantial structure built on a quicksand foundation of unprocessed, generally incomplete, and qualitative rather than quantitative data. What can be done?

My considered recommendation is that in this and other countries a nucleus of deeply interested scientists and engineers should establish on a modest but continuing basis a loosely-knit "institute" for the study of the UFO phenomenon. The scope, diversity, and extent of the work of each institute would be set by the funds and time available. Of course, funds always remain a problem and would have to be solicited locally from private sources or, in some cases, from governments and scientific associations. A great deal can be done even with modest research grants if they are properly administered.

Since the phenomenon is global, contact between groups in various countries must be maintained, and some form of communication is needed, perhaps eventually growing into an international journal devoted to this study.

I would also strongly recommend that a member country of the United Nations propose in the General Assembly that a committee be set up within the United Nations structure to aid and facilitate communications between these small groups of scientists in various countries.* Such a committee would not, of course, commit the United Nations either to financial or directive support but would be, in effect, a clearing house for the exchange of information. In this sense it would act as many already existing scientific unions (for example, the International Astronomical Union) operate. They provide a means whereby specialists in various countries are made known to

* On June 18, 1966, U Thant, then Secretary General of the United Nations, expressed to Mr. John Fuller and me his strong interest in the UFO problem. During an hour-long discussion with us he pointed out the similar concern that had been expressed to him by General Assembly members from several countries. He told us that he was sympathetic to UN action but that UN action would have to be initiated by a member nation.

each other and can communicate and plan mutual programs without the lag of formal publication. The International Astronomical Union, for instance, has more than forty commissions, each covering specific facets of astronomy, and each facilitates communication between astronomers pursuing their particular astronomical specialty.

In a similar manner, there is need for specialization in studies of the UFO phenomenon. Progress comes through specialization; what serious workers there have been so far in the UFO field could, in general, be likened to general practitioners in medicine. Although it may seem far-fetched to the reader, there is ample room in the study of the UFO phenomenon for specialists in the same sense that in the medical field we have heart specialists, pediatricians, gynecologists, and so forth. Phillips, for instance, has specialized in the study of ground markings reportedly made by UFO landings. Similar specialized work is needed in reported cases of interference with ignition systems on automobiles, UFO effects on animals, trajectories and kinematics of UFO flight, morphology of humanoids, reported communications with occupants, the spectral characteristics of nocturnal lights, and many more aspects of the UFO phenomenon. One can indeed envision occasional international meetings (as is done every three years in the case of the Astronomical Union) during which such specialists can meet and report their findings. The particular programs of serious investigation would, of course, be chosen by the cooperating scientists. Were I responsible for such planning, I would first divide the total program into two major parts; they might be called, respectively, the *active* and the *passive*.

The objectives of the *active* program would be to obtain quantitative observations of the UFO phenomenon itself. Ideally, this would involve being present at the time of a sighting, equipped with cameras, spectrograph, tape recorder, Geiger counter, infrared equipment, surveying equipment, etc. in order to get movies of the event, photographs of the UFO forms, spectrograms to determine whether the radiation was comprised of continuous radiation or emission and absorption lines, accurate triangulation fixes to determine distances, and accurate measurements of landing marks, broken trees, etc. In short, such specialists would document quantitatively the event as it was actually happening.

However, this ideal could be attained only by accident. The occurrence of a UFO Close Encounter from all accounts is as unpredictable as the landing of a meteorite, and the chances of obtaining quantitative measures of that event—for example, movies of a meteorite landing—are indeed slight. I know of no astronomer, for example, who has ever observed the actual landing of a meteorite. (Meteor flashes in the sky, of course, are quite a different thing—I speak here of the landing of a physical object.)

The comparison is apt, for there was a time when the existence of meteorites was denounced by official science, and stories of their fall, told by reputable witnesses, were regarded as "old wives' tales" simply because it seemed preposterous that stones could fall from the sky. In 1801 Thomas Jefferson was reported to have said that he would sooner believe that two Yankee professors had lied than that stones had fallen from the sky.

Let us suppose, however, that Thomas Jefferson had set up a Ben Franklin committee to settle the question of whether stones did indeed fall from the sky. If Ben Franklin had decided to set up meteorite landing observation stations around the country, the cost would have been prohibitive and the results, barring a most fortunate accident, would have been nil. Even had photography existed in those days, the cost of establishing photographic stations every few hundred yards across the United States would have, of course, been completely out of the question. The chances are that such an active meteorite program would surely have come to naught. Similarly, also barring a very happy accident, setting up comprehensive UFO observing posts all around the world (for we must remember the phenomenon is global) would be totally prohibitive in cost and at any rate might well yield nothing.

It is often reported that UFO sightings seem to cluster in "UFO hot" areas. If this effect is not due to publicity, hysteria, and general suggestibility (and, of course, standard screening of the reports would establish quickly whether the successive reports qualified as UFO reports), observing equipment rushed to that area might greatly increase the probability of obtaining firsthand data. It would appear that a true clustering of UFO reports does sometimes occur (Oklahoma in 1965, rural France in 1954, Argentina in 1963) and that therefore some hope lies in this direction.

The *passive* part of the program would be, of course, the careful statistical study of the data as already outlined. A bridge between the two phases would be the active investigation of fairly recent cases, in which the reported object has long gone but has left its traces on the ground, on plants, and, of course, in the memories of the observers. The active collection of data before a case is too old is of paramount importance.

This aspect requires the availability of thoroughly trained investigators, and urgency requires that their job be full time when the occasion demands. And this requires adequate funds.

If funds were no object(!) and I were directing a UFO institute, I would personally train an adequate number of fulltime investigators and then, when a particularly interesting UFO report came along, assign two investigators to bird-dog the case until every bit of potentially available data was obtained. This might take a week, a month, six months, or even longer. It would make no difference; this would be their fulltime job until every lead, every clue, every available witness had been explored and every possible measurement made.

The present poor state of UFO data has come about because first, the original reporter usually does not know what precise data are needed; and second, the investigators have done their work as a hobby, on weekends or whenever spare time was available, and too often they have lacked training in garnering the relevant data. The active phase of an institute's program would thus be in effect on an on-call basis, and the passive aspect would be a continuing program of data reduction.

A great wealth of data, highly variable in quality, has been garnered over the past two decades. In its present form it is much akin to low grade ore, which must be processed and refined before it is of value. Or perhaps a more appropriate analogy might be the case of Marie Curie and the tons of pitchblende that she had to process before they yielded a mite of radium. Those of us who have spent time on the UFO problem are convinced that the probability is very high that there is "radium in the pitchblende" in the quantities of reports. Although it will be a Herculean task to cull and refine existing UFO data, I feel a rich reward awaits a person or a group that assumes this task with dedication. For if there is indeed paydirt in the ore of UFO data, it might well represent a scientific breakthrough of major magnitude. It might call for reassignment and rearrangement of many of our established concepts of the physical world, far greater even than the rearrangements that were necessary when relativity and quantum mechanics demanded entrance into our formerly cozy picture of the world.

Obtaining the cooperation of the various UFO organizations around the world in making their files available for a major statistical study is essential to its full success. Whereas current cases can be studied locally, a major statistical study can be truly meaningful only if universal data garnered in the past are used. Clearly this would require that the work be done by an organization meriting the respect of individual organizations in various countries; this I believe could far more easily come about if the worldwide effort had the sponsorship of an international scientific union or of the United Nations. In the United States the private organizations APRO and NICAP would need absolute assurance that their cooperation would not be treated in the cavalier manner displayed by the Condon committee.

The Blue Book files are, according to law, unclassified and available to legitimate scientific investigators. The files of Great Britain, France, Australia, and many other countries, both official and private, represent a potential source of valuable data but may be subject to various security regulations. It is my understanding that the British military files of UFO reports cannot be made public until a period of thirty years has elapsed.

Nonetheless, access to all the data that exist is not necessary for a valid statistical study. Whatever does become available, however, must be incorporated into a homogeneous format. Many groups and individuals of differing experience in data processing and in UFO investigation are at the present endeavoring to put their material in machine-readable form. While this is a most laudable intent, unless their coding is mutually compatible, the blending of worldwide data will come to naught or will eventually require redoing in a uniform code. As soon as possible, international agreement of the method of coding UFO data is necessary; this could well be a primary function of a United Nations-sponsored committee.

Proper computerization of the data is absolutely essential in seeking patterns in UFO behavior, in establishing cross-correlations, and in seeking possible differences or similarities in behavior in different countries. This is not mere cataloging and busy work. The modern computer used with appropriate software (a sophisticated nonprocedural language) can establish meaningful correlations if they exist. For example, of the hundreds of cases of reported automobile failure in the presence of a UFO, what do these cases have in common? In what ways do they differ? What failed first—the radio, the lights, the motor? And when a UFO exhibits a sequence of colors, what is the most frequent color, the most frequent sequence?

Such analysis, coupled with the *active* program of on-the-spot investigations of a truly scientific character, should accomplish the first objective of a positive UFO program: to establish the reality of the UFO as a legitimate subject for further scientific study. If definite patterns and other correlations can be established for UFOs reported in many different countries by people with different levels of culture, the probability that such correlations happened by chance as a result of random misperceptions would be vanishingly small. The probability, therefore, that the UFO represents something truly new in science—new empirical observations—would be a virtual certainty.

EPILOGUE

Beyond the Blue Book Horizon

It is a capital mistake to theorize before you have all the evidence. It biases the judgment.

—SHERLOCK HOLMES, *A STUDY IN SCARLET*

Holmes surely exaggerated, for one never has all the evidence. In any investigation, however, there generally comes a time in which the investigators feel that there is a sufficient body of evidence to theorize productively, especially in suggesting leads for further investigation. The results of such theories, in turn, stimulate further theorizing.

In the UFO problem, however, much more quantitative evidence is needed before theorizing is likely to be productive. After more than twenty years' association with the problem, I still have few answers and no viable hypothesis. And I have no desire to act the prophet.

I say *association* rather than study, for during the first several years of that association I felt, as did virtually all my colleagues, that the subject was nonsensical, and I had little inclination to give it serious study. Later, as it became increasingly clear to me that the subject did merit study, I had no funds, no mechanism, and certainly little time with the press of professional duties to undertake the kind of study that would have been comprehensive enough to be significant.

My consulting work with the Air Force most emphatically did not provide such a mechanism, although it did provide me with data for possible future study. Therefore, when the Condon committee was formed, even though I knew I was not to be a member, I applauded the move with hope and naive anticipation. I recognized that the funds provided were insufficient for a full-scale attack on the problem, but I felt that if the funds were spent wisely and the investigation conducted without prejudice and in a true scientific spirit, the merit of a more extensive and continuing study would be demonstrated. Indeed, a close reading of the report and its puzzling cases has provided this demonstration. Several scientists told me that it was a study of the Condon Report that first led them to realize that the UFO problem was one worthy of investigation. But the story of the Condon committee has been told. The repudiation of its summary conclusions awaits, in my opinion, only a calm and unbiased study of the UFO phenomenon, a study that will organize, refine, and order the evidence; only then can we profitably entertain and test hypotheses.

It would be silly to pretend, however, that explanations for the UFO phenomenon—possible ones as well as highly fanciful explanations—have not already been presented. Indeed, a goodly part of the enthusiast literature is devoted to their exposition or to uncritical acceptance of a particular hypothesis—most frequently, of course, the extraterrestrial hypothesis.

The serious investigator should resist the temptation to theorize prematurely, especially in this instance, for this subject is beset by a number of difficulties not normally encountered elsewhere in scientific research. For example, in a typical frontline research topic such as elementary particle physics, each new piece of experimental data is immediately confronted with a multiplicity of hypotheses from the theorists. They are safe in proposing theories on the basis of scanty new evidence (long before all the evidence is available) because they are operating well within the bounds of a recognized and accepted framework of physical concepts. Indeed, many theorists rush to develop theoretical models of the system of interest with only casual regard to their empirical verification. For them theory-making is a professional game, intended to stir the experimentalist to devise new experiments to prove or disprove the theory. In either case the theorist is happy. What theoretical astrophysicist, for instance, would consider waiting until every pulsar had been thoroughly cataloged and studied before embarking on speculations concerning neutron stars?

Sometimes it happens that theory long precedes any empirical observation. It was once theorized by an astronomer that the moon's surface was covered with so deep a layer of dust that ships from earth might sink out of sight. When Apollo gave the lie to that particular theory, did its originator hang his head in shame? Not at all! He went on with many new theories, some of them proving correct. As one of the ablest astrophysicists of our time, he knew that theory-spinning is not only fun but that it, especially if controversial enough, can be a very sharp spur to action.

Unfortunately there are several problems concerning UFOs that advise against such uninhibited theorizing. The first is philosophical. The scientific tradition since the time of Galileo has evolved a logical and methodological structure that has proved highly successful in allowing us to understand a wide class of phenomena. This tradition is sacrosanct among the scientific community simply because it has worked with outstanding success. In this procedure it is usual for the experimentalist to try to manipulate the environment in such a way that the significant aspects of the phenomenon are isolated from the irrelevant and spurious. To put it another way, he devises means of separating the signal from the noise. In this way precise causal relationships between quantities and items suggest themselves; in the case of the physical sciences such relationships are often expressed in mathematical form.

Even when active experimentation in the laboratory is precluded, as in astronomy, the astronomer can still extract the signal from the noise by the use of special instrumentation when the phenomenon (such as an eclipse) becomes available to him. Relationships between certain parameters associated with the phenomenon thus become apparent, and further testing and experimentation can then establish them beyond all reasonable doubt. They become a scientific fact. If it were not possible to operate in the above fashion, science would be immensely more difficult; it would be virtually impossible to extract and separate the fundamental elements inherent in the phenomenon from the host of irrelevant and coincidental things always present in an observed situation.

This is the situation in the case of the UFOs, which are totally beyond experimental control. Moreover, they are transient, unscheduled, and obtrude upon an observer who is often not competent to make a dispassionate analysis of the totally unexpected and surprising situation. Consequently, the significant features of the phenomenon may be buried among incidental—but much more conspicuous—features emphasized in the reports.

UFOs, however, are not alone in this category; they share these particular difficulties with many other phenomena, such as ball lightning and meteorites, for which one must rely on the fortuitous observations of the layman for one's data. That is why subjects such as these gained scientific respectability so slowly, particularly when an explanation was hard to find because the phenomenon did not fit the scientific framework of the moment.

We may even have to face the fact that the scientific framework, by its very internal logic, excludes certain classes of phenomena, of which UFOs may be one. One of the most exasperating and even repugnant features of the subject is its apparent irrationality. However, as our concept of rationality is a by-product of the scientifically oriented society in which we live, it should not surprise us if a phenomenon that is inaccessible to scientific procedure appears irrational.

It is just here that we encounter a second difficulty of the UFO problem. It cannot, at least at present, be separated from the social condition in which it is embedded. We are accustomed to the almost complete isolation of the behavioral sciences from the physical sciences, yet in this problem we have a situation in which the two are inextricably mixed. Whether or not separate, nontrivial physical and behavioral components will emerge as serious study by both disciplines remains to be seen, but it would be premature to reduce the importance of either.

Of course, this discussion would be unnecessary if an obvious explanation of UFOs was at hand. In one's frustration it is all too easy to seize on an explanation of the "men from Mars" variety and to ignore the many UFO features unaccounted for. But to do this is to fall into the very trap we have just discussed. We may be inadvertently and artificially increasing the significance of the conspicuous features while the part we ignore—or that which is not reported by the untrained witnesses—may contain the clue to the whole subject.

What needs to be explained has been amply outlined in the descriptions of the six basic UFO observational prototypes given in chapters six through eleven. The most persistent and enigmatic features seem to be the localization of the phenomenon in space and time, its apparently intelligent characteristics (of a rather puerile kind), its appearance of operating outside the established laws of physics, and its peculiar preferences for certain situations. The frequently reported presence of humanoids capable of moving about in comfort in our highly restrictive terrestrial environment, and their association with craft, exhibiting at times near-zero inertial mass yet able to leave physical traces of their presence, is surely a phenomenon beyond the pale of mid-twentieth century physics. But there will surely be, we hope, a twenty-first century science and a thirtieth century science, and perhaps they will encompass the UFO phenomenon as twentieth century science has encompassed the aurora borealis, a feat unimaginable to nineteenth century science, which likewise was incapable of explaining how the sun and stars shine.

We work in the brilliant spotlight of the present, only dimly conscious of the penumbra of the past and quite unable to illuminate the darkness of the future. Let us imagine for a moment a covered wagon train of not much more than a century ago, winding its long journey to the west. It is encamped for the night, its wagons in a circle, sentries posted, and the travelers gathered about a campfire for warmth and cheer. Some speak of the future, but they speak, as they must, with the words and concepts of their day. But even were they inspired by some kindly muse of the future to speak of making their entire journey in a matter of hours, flying through the air, and of watching scenes by television and hearing voices speaking on another continent, this gifted one could not have put into words a glimmer of how these wondrous things might be accomplished. The vocabulary for such descriptions—electrons, transistors, integrated circuits, jet engines—the jargon vehicle of technical communications would not yet exist for yet a century. They would be helplessly incoherent for want of words as vehicles for their thoughts.

Would one care to venture a guess at the technical vocabulary of the year 373,475 (assuming intelligent life still exists on earth) and to predict the concepts and knowledge for which it will be a vehicle?

Does such an advanced knowledge and technology already exist somewhere in space? The sun, our parent star, is but one star out of billions in our galaxy, and our galaxy is but one of many millions, each with its billions of stars. It is statistically improbable that our sun is the only star out of quadrillions of stars to have planets. That would be somewhat like claiming that acorns can be found lying near only one oak tree in the world.

Even if we limit our thinking to the billions of stars in our galaxy alone, we know that our galaxy was in existence for billions of years before our sun appeared. Thus the stage was set long ago for this possibility, the possibility of civilizations as greatly advanced beyond us as we are beyond mice. For instance, Fred Hoyle[1] has conjectured that it is possible that a great intragalactic communications network exists but that we are like a settler in the wilderness who as yet has no telephone.

Such ideas, once forbidding and even revolting to our geocentric minds, no longer shock us as we slowly grow out of our cosmic provincialism. Such concepts, however, have little to do directly with our problem at the moment save that they present one possible hypothesis for study. But talk of extraterrestrial visitors or the more esoteric notions of time travel or of parallel universes is as inappropriate as the mass hallucination hypothesis for UFOs at this stage. Kuhn has commented that scientific progress tends to be revolutionary rather than evolutionary, and the above concepts are, despite their bizarre nature, merely imaginative extensions of current concepts. When the long-awaited solution to the UFO problem comes, I believe that it will prove to be not merely the next small step in the march of science but a mighty and totally unexpected quantum jump.

APPENDIX 1

Descriptions of Sightings Discussed in Text

DAYLIGHT DISCS

CASE	DATE	TIME	LOCATION	NO. OBSERVERS	DURATION	S	P
DD-1	Jan 15, 1968	7:25 p.m.	Three Hills, AL	2+2	10 min	2	4
DD-2	Apr 11, 1964	6:30 p.m.	Homer, NY	3	45 min	3	8
DD-3	Oct 21, 1967	6:16 a.m.	Blytheville AFB, AR	3	30 sec	2	8
DD-4	Mar 24, 1967	8:45 a.m.	Los Alamos, NM	1+1	30 sec	1	8
DD-5	Aug 15, 1950	11:30 a.m.	Great Falls, MT	2	1 min	2	9
DD-6	Jul 3, 1967	5:30 p.m.	SW of Calgary, AL	3	25 sec	3	7
DD-7	Jun 28, 1952	1:20 p.m.	Kirtland AFB, Albuquerque, NM	2	30 sec	2	4
DD-8	Jan 16, 1952	Afternoon	Artesia, NM	2+4	40 sec	2	6
DD-9	Oct 15, 1953	10:10 a.m.	Minneapolis, MN	3	40 sec	2	9
DD-10	Apr 24, 1949	10:30 a.m.	White Sands, NM	5	1 min	2	9
DD-11	Jan 30, 1967	8:04 a.m.	Crosby, ND	5+2	40 sec	2	5
DD-12	Mar 26, 1967	4:00 p.m.	New Winchester, OH	2+3	5 min	1	5
DD-13	Apr 1, 1967	11:30 a.m.	Kenosha Pass, CO	2	2 min	1	5
DD-14	Aug 9, 1965	6:00 p.m.	Long Island, NY	5	5 min	1	8
DD-15	Sept 1965	4:00 p.m.	Fort Sill, OK	15	10 min	1	7

NOCTURNAL LIGHTS

CASE	DATE	TIME	LOCATION	NO. OBSERVERS	DURATION	S	P
NL-1	Jan 14, 1966	5:55 p.m.	Weston, MA	4	14 min	2	9
NL-2	Nov 26, 1968	5:40 p.m.	Bismarck, ND	4+1+1	5–7 min	2	9
NL-3	Sep 17, 1968	1:00 a.m.	Nellis AFB, NV	2	40 min	1	6
NL-4	Dec 17, 1966	Late evening	Whittier, CA	2	5 min	2	6
NL-5	Spring 1961	Late evening	Millville, NJ	2	7 min	2	4
NL-6	Oct 20, 1966	11:50 p.m.	Moose Jaw, Sask	2	4 min	2	6
NL-7	Dec 24, 1967	8:30 p.m.	Belmont, MA	8	15–20 min	2	9
NL-8	May 14, 1970	9:45 p.m.	Bangor, ME	2	2–3 min	2	9
NL-9	Sep 22, 1966	3:00 a.m.	Deadwood, SD	2+2	1 hr +	2	7
NL-10	Aug 18, 1964	12:35 p.m.	Atlantic Ocean, 200 mi E Dover, DE	4	2 min	1	8
NL-11	Summer 1960	2:00 a.m.	Walkerton, Ont	3+2	1 hr +	3	8
NL-12	Feb 25, 1967	7:50 p.m.	Fargo, ND	2	4 min	2	8
NL-13	Mar 9, 1967	9:10 p.m.	Onawa, IA	3	1 min +	1	5

RADAR-VISUAL

CASE	DATE	TIME	LOCATION	NO. OBSERVERS PLUS RADAR	DURATION	S	P
RV-1	Jan 13, 1967	10:00 p.m.	New Winslow, AZ	3	25 min	2	8
RV-2	May 4, 1966	4:30 p.m.	Charleston, WV	3	5 min	2	8
RV-3	Nov 4, 1957	10:45 p.m.	Kirtland AFB, Albur-querque, NM		5 min vis. 20 min radar	3	8
RV-4	Aug 13–14, 1956	10:30 p.m. to 3:30 a.m.	Lakenheath, Eng	1+1+1+1	5 hr	3	8
RV-5	May 6, 1965	1:10 a.m.	Philippine Sea	12	8 min	2	8
RV-6	Jun 3, 1957	9:35 p.m.	Near Shreveport, LA	3+1	1 hr	1	5
RV-7	Feb 13, 1957	2:30 a.m.	Lincoln, AFB, NE	5	25 min	2	7
RV-8	Jul 17, 1957	4:10 p.m.	SW United States	6	1½ hr	4	9
RV-9	May 16, 1967	10:10 p.m.	Aboard SS *Point Sur*, Gulf of Mexico	4	50 min	3	8
RV-10	Dec 6, 1952	5:25 a.m.	Gulf of Mexico	6	10 min	2	8

CLOSE ENCOUNTERS OF THE FIRST KIND

CASE	DATE	TIME	LOCATION	NO. OBSERVERS	DURATION	S	P
CE I-1	Oct 27, 1967	3:05 a.m.	Parshall, ND	1+1	5 min	2	6
CE I-2	Jan 11, 1966	7:40 p.m.	Meyerstown, PA	4	10 min	4	6
CE I-3	Feb 6, 1966	6:05 a.m.	Nederland, TX	3	5 min	4	4
CE I-4	Jun 19, 1965	4:00 a.m.	Rocky, OK	2	2-3 min	4	7
CE I-5	Apr 22, 1966	9:00 p.m.	Beverly, MA	10	30 min	4	8
CE I-6	Jul 22, 1966	11:30 p.m.	Freemont, IN	2	5-8 min	5	5
CE I-7	Apr 17, 1967	9:00 p.m.	Jefferson City, MO	1+1+2	10-15 min	3	8
CE I-8	Aug 20, 1955	10:45 p.m.	Kenora, Ont	2	4 min	2	8
CE I-9	Apr 17, 1966	5:05 a.m.	Portage County, OH	2+1+1	1 hr, 35 min	4	8
CE I-10	Apr 3, 1964	9:00 p.m.	Monticello, WI	3	5-10 min	3	9
CE I-11	Mar 8, 1965	7:40 p.m.	Mt. Airy, MD	3	3 min	3	8
CE I-12	Oct 10, 1966	5:20 p.m.	Newton, IL	6	3-4 min	4	9
CE I-13	Jun 26, 1963	1:00 a.m.	Weymouth, MA	2	1 min	2	5
CE I-14	Oct 14, 1967	2:30 a.m.	Mendota, CA	3	3 min	2	8

CLOSE ENCOUNTERS OF THE SECOND KIND

CASE	DATE	TIME	LOCATION	NO. OBSERVERS	DURATION	S	P
CE II-1	Nov 2, 1957	11:00 p.m.	Vicinity of Level-land, TX	2+1+1+1+1+1 +1+1+2+1	2½ hr	5	8
CE II-2	Apr 3, 1968	8:10 p.m.	Cochrane, WI	2	5-10 min	4	8
CE II-3	Aug 1954	6:00 p.m.	Tananarive, Madagascar	2,000	2 min	4	7
CE II-4	Oct 26, 1958	10:30 p.m.	Loch Raven Dam, DE	2	1-2 min	3	8
CE II-5	Mar 8, 1967	1:05 a.m.	Leominster, MA	2	4 min	3	7
CE II-6	Jan 23, 1965	8:40 a.m.	Williamsburg, VA	1+1	2 min	3	5
CE II-7	Aug 4, 1968	4:15 a.m.	Regina, Sask	3	10 min	4	4
CE II-8	Jan 20, 1967	6:50 p.m.	Methuen, MA	3+3	15 min	3	8
CE II-9	Aug 29, 1965	7:00 p.m.	Cherry Creek, NY	4	2-3 min	3	7
CE II-10	Dec 8, 1957	5:30 p.m.	NW Okinawa	3	3 min	5	6
CE II-11	Sep 3, 1965	11:00 p.m.	Damon, TX	2	5-20 min	3	8
CE II-12	Jul 12, 1969	11:00 p.m.	Van Horne, IA	2+1	2 min	4	8
CE II-13	Apr 21, 1967	1:00 a.m.	Ephrata, WA	4	5 min	2	5

CLOSE ENCOUNTERS OF THE SECOND KIND *CONTINUED*

CASE	DATE	TIME	LOCATION	NO. OBSERVERS	DURATION	S	P
CE II-14	Apr 1, 1968	10:45 p.m.	Frankfort, KY	2	5 min	2	4
CE II-15	Aug 22, 1957	9:30 p.m.	Cecil NAS, FL	2	40 min	3	4
CE II-16	Apr 14, 1957	3:00 p.m.	Vins, France	2+1	1-2 min	3	5
CE II-17	May 29, 1968	10:00 p.m.	Mosinee, WI	6	2-3 min	2	4
CE II-18	Jun 18, 1967	11:00 p.m.	Falcon Lake, Ont	6	30 min	3	5
CE II-19	Oct 7, 1966	8:30 p.m.	Indian Lake, MI	14	1 hr	4	7
CE II-20	Jun 1957	10:30 p.m.	Warrensburg, MO	3	45 min	4	7
CE II-21	Jan 12, 1965	8:20 p.m.	Custer, WA	4+1	5 min	4	6
CE II-22	May 11, 1969	2:00 a.m.	Chapeau, Que	1+4	5 min	4	5
CE II-23	Oct 11, 1967	Late evening	Aldersyde, AL	2	2-4 min	3	5

CLOSE ENCOUNTERS OF THE THIRD KIND

CASE	DATE	TIME	LOCATION	NO. OBSERVERS	DURATION	S	P
CE III-1	Apr 24, 1964	5:45 p.m.	Socorro, NM	1+1	5-10 min	5	6
CE III-2	Nov 1961	Past midnight	Minot, ND	4	1 hr	5	5
CE III-3	Jun 26, 1958	Dusk	Boianai, New Guinea	25+	3 hr	5	8
CE III-4	Aug 21, 1955	Evening	Kelly-Hopkinsville, KY	7	4 hr	5	5
CE III-5	Sep, 19, 1961	Near midnight	Whitfield, NH	2	1 hr	5	4

APPENDIX 2

Analysis of the Papua-Father Gill Case by Donald H. Menzel

In this spectacular case Father Gill and a great many uneducated natives of Papua reported seeing some remarkable objects in the sky. Most of the sightings occurred in the early evening, shortly after sunset. I find it significant that Venus was a very conspicuous object, setting about three hours after the sun. It reached greatest elongation East on June 23 and attained maximum brilliancy on July 26.

I think it significant that, despite the brilliance of Venus, none of the sightings by Father Gill and the Mission group refers to that planet. Two officers recognized that Venus "could be expected to be seen from this station in approximately the same direction as the bright light was first seen." He states that he saw the planet Venus but he had the opinion that the object seen by the Mission group was lower than Venus and more to the North. This is an expression of opinion, however, rather than a definite observation. Robert L. Smith, Cadet Patrol Officer, saw Venus in the early evening of July 6, but he apparently did not see any UFO. He mentions looking considerably after midnight, and seeing a bright object, which almost certainly was the planet Jupiter. He also saw some "shooting stars."

Many of the experts say that the UFO "looked like a star." However, there remained to be explained the remarkable gyrations reported principally by Father Gill. I find unconvincing the fact that a number of the mission boys and girls seemed to corroborate the sighting.

The following could have been the explanation, and, in fact, some experiments I have performed indicate that it probably was the correct explanation. Some of these could still be checked. We are first to assume that Father Gill and Stephen Gill Moi (teacher) both suffer from appreciable myopia and that they were not wearing spectacles during the sighting. They probably had appreciable stigmatism as well, so that the image of Venus was large and definitely elongated. Something of this sort is necessary to account for the difference in appearance of the UFO as reported by the two individuals. Father Gill had the long access of the vehicle horizontal; Stephen had it more nearly vertical. The human eye executes erratic motions, which make an object such as a star or planet appear to be vibrating when, in fact, the object is standing still. Atmospheric effects account for the rapid changes in color.

But what about the reported men waving? Could this have been an illusion? With a myopic eye, the excursions of the eyelid over the pupil perform a sort of optical knife edge. Out-of-focus nature of the eyelashes and out-of-focus images of the eyelashes and a defraction resulting from squinting as a nearsighted person tries to improve his vision. The waving to the occupants and the reported waving back might not have been as universally observed as Father Gill thought. He reported gasps of either joy or surprise, perhaps both. Could these gasps have been of incredulity because of the inability to see what Father Gill was reporting? After all, in a Mission of this sort, the natives must have been conditioned to miracles and the like.

To simulate this phenomenon I secured a positive spectacle lens of about four diopters strength. I intend to repeat the experiment with a lens having an appreciable stigmatism, to simulate the assumed myopic condition of Father Gill. Then by blinking I could readily imagine some of the phenomena that he reported. Part of the effect may have been from eyelid defraction apart from irregularities, such as blood cells on the retina. These would most certainly show up under the circumstances. Father Gill simply assumed that the other people were seeing what he saw. Although a great many witnesses signed the report, I doubt very much that they knew what they were signing or why. They would certainly have been mystified as to why their great leader was seeing something that was invisible to them. On the other hand, they would not have been too surprised because after all, they looked upon Father Gill as a holy man. Many people in this world need glasses and fail to wear them. I should be very much interested to know whether or not Father Gill wears glasses, what his correction is, and finally, whether he was wearing them on that evening. Since a very simple hypothesis accounts, without any strain, for the reported observations, I shall henceforth consider the Father Gill case as solved. Moreover, I feel that the same phenomena are responsible for some of the more spectacular, unsolved cases in the Air Force files.

DECEMBER 20, 1967

APPENDIX 3

Letter of Resignation from Mary Louise Armstrong to Doctor Edward Condon

24 February 1968
Dr. Edward U. Condon, Director UFO Project
University of Colorado
Boulder, Colorado 80302

Dear Dr. Condon:

This letter shall be a written presentation of the points we discussed Thursday morning, 22 February 1968.

Since it is apparent to the staff of the UFO project, as well as to you, that we are in a real dilemma over the disagreement and low morale within the study as a result of the last two weeks, I feel it is necessary to examine what, in my opinion, has been the primary cause of the problems that exist. I sincerely hope that the project will continue on a very different basis than before, that communication between you and your staff will improve greatly, and that what we all want out of the study will occur; that is, a final report that everybody can be satisfied with.

It is my belief that all of the project members to a certain degree must share in the responsibility for the present situation, if for no other reason than that we haven't come to you sooner about our misgivings. However, I strongly believe that, had Bob not been the individual who *directly* and on a day-to-day basis administered the project, we would not be in this situation. I think there is an almost unanimous "lack of confidence" in him as the project coordinator and in his exercise of the power of that position. (I must emphasize at the outset that I realize each person must represent only his—or her—opinions and that when I refer to other staff members I only state my observations of their dissatisfaction.)

Listed below are my reasons and a discussion of them as to why I think Bob is responsible for the conflict and why, in my opinion, had *you* handled the direction of our activities, there would not have been such a serious conflict.

Bob's attitude from the beginning has been one of negativism. While I doubt that *he* would agree with this statement, I would expect most of the staff would. Bob showed little interest in keeping current on sightings, either by reading or talking with those who did. At one point in our study, it was agreed that a certain number of the staff would read a designated group of reports systematically and then meet to go over what they had read. In this way it was hoped that some meaningful discussion would be stimulated as to what could be said, if anything, about the reports. Saunders carefully set aside reports on a check-out basis, so that everyone on the committee would have a chance to read them. Bob checked some out, but, to my knowledge, never really read them, and certainly never encouraged the proposed discussions to actually take place. I think he, as project coordinator, should have taken the initiative to see that this program was carried out. Moreover, much of what I want to discuss later concerning Bob's premature writing of the final report at this time deals directly with what *can* or *cannot* be said about sighting reports. *To me, too much of his time has been spent in worrying about what kinds of "language" should be used in the final report so as to most cleverly avoid having to say anything definitive about the UFO problem.* Very little time, on the other hand, has been spent in reviewing the data on which he might base his conclusions.

Bob complained to me once not long ago that he was supposed to be part of the committee that would meet to decide which sightings should be investigated by our field teams, but that he had *not* been contacted when it was time to make these decisions. I asked Norman if this was true and he categorically denied it. He stated that Bob had been consulted every time and, for the most part, had declined to take part. However, even if Norman had not contacted him, geographically Bob was close enough to the situation (which you were not) to participate, if that was what he really wanted, in any dialogue that any of the rest of the staff could complain of not being included in any decision-making process. Certainly it was *Bob's* responsibility to take the initiative. After all, right or wrong, he, as the project coordinator, had it in his power at any time to change the procedure.

This raises the question of what Bob actually has done with his time. I feel much of it has been meaningless and apart from what should have concerned our study, given the time and budgetary limitations.

Bob has traveled a lot. I realize that many of these trips concerned subjects that were relevant to the UFO problem—relevant in the way that the staff envisions "relevance"—i.e., running down information on the Heflin case, two current sighting investigations (very early in the project), and visits to SRI, Rand, Hippler, and Ratchford. However, many of the trips seemed to me to deal with unimportant aspects of the UFO problem. Bob has given quite a few speeches (which ostensibly was not to be one of our project's responsibilities). Some of them include the Boeing Corporation in Seattle, The Rand Corporation in Santa Monica, the American Meteorological Society in Colorado Springs, and the IEEE in Los Angeles. He has justified this "speaking tour" as being educational or university-associated or as dealing with scientific institutions. Concerning the travel aspect, however, I feel the biggest misuse of his travel time was his trip to Europe. Granted there is a justification for someone going to Europe (or South America, Africa, or anywhere outside the US) to see what the UFO situation is internationally. On any trip to Europe I would think a visit to Michel and Bowen would have been appropriate, if not compulsory. However, visits with the Ministry of Defence in England, and Swedish Defense Group, Loch Ness, and a man named Erich Halik in Vienna (who, as far as I can tell, only represents one of a large group of people from whom we get letters every day suggesting how to build "flying saucers," solve propulsion system problems, etc.) seem to be remote from the problem of UFOs, if not altogether irrelevant, and out of the scope of what our project can accomplish with limited time. In addition, though Bob has discussed his European trip with the staff, I have never seen a written trip report. In the past he has been the one who has insisted on documentation of every trip we have taken.

It can be argued, and reasonably, that Bob has had to deal with many of the straight administrative problems (i.e., finances, subcontracts, organization of the office and jobs that individuals would be doing) and that he has contributed to seeing that that kind of work gets done. Moreover, it's true that the staff was given a free hand to do just as they wished. At the same time, however, Bob initiated a good many individual projects, but did not follow through on them to any great extent, or even keep abreast of what others were doing. If he had, I do not believe that he could have justified the writing of his thoughts as conclusions for the final report when, not only is it not *his* report and he is not the Director, but he did not consult the people who have essentially done all the work with the data. Why is it that Craig, Saunders, Levine, Wadsworth, Ahrens and others have all arrived at such radically different conclusions from Bob's? It is not my impression that they came into the project with any particular bias concerning the UFO problem. I think that there is a fairly good concensus among the team members that there is enough data in the UFO question to warrant further study. This is not to say, as no one of us would, that we are definitely being visited by vehicles from outer space. But to say in our final report, as I believe Bob would like to, that although we can't prove "ETI" does not exist, we can say that there isn't much evidence to suggest it does, would not be correct. I do not understand how he can make such a statement when those who have done the work of digging into the sighting information do not think this is true. A dialogue will have to occur eventually in which both sides of the question are debated within the group, but to be putting these ideas down on paper in the form of conclusions and discussing them with people outside the project is presumptious and wrong.

In the memorandum Bob wrote to David Williamson of NASA on 12 December 1967 he states:

1. "In the absence of scientific data, our answer is probably going to be that (aerial phenomena of unknown origin (UFOs) that represent

phenomena or stimuli outside the range of present-day scientific knowledge) it is possible but that there is nothing to support an assertion that it's true . . .

2. "The second part of the letter (Dolittle's letter to J. T. Ratchford of 2 August 1967) sets up the requirement for the technical side of the study. It provides that the current state of knowledge in the physical, behavioral, and social sciences be brought to bear on the public policy objective. The point here is that it is our job to do the science (but of course our finding is that, because there are no data, we can't do any proper physical science); it is the Air Force's responsibility to apply the scientific findings to the public policy decisions . . .
3. "We let the Air Force off the hook, and we shouldn't, if we do other than say flatly that, using all the tools of science, we have not been able to reach any solution of the UFO problem."

The first statement raises the question of the impossibility of using science in the study of UFOs. I would think most of the staff would certainly take strong issue with that. The second statement appears to say that it is not our job or responsibility to make recommendations on the UFO question, but only to review the problem scientifically and submit it to the National Academy of Sciences. I would agree that, seen in the strictest interpretation of the contract and Dolittle's letter, that could be correct. But who of us does not feel that this is primarily a question of public responsibility and that we very definitely do have to make recommendations, at least in the sense that the UFO problem does or does not warrant further study. The third statement gives the impression that we have to reach a "solution," and that if we don't answer positively or negatively the question of ETI, we have not reached a "solution." I would think that the word "solution" means a very different thing to Bob than it does to the staff and possibly to you, too.

The very fact that Bob has discussed so freely the UFO study with people such as Williamson, Asimov, Branscomb, Higman (and others) and, while I do not feel there is anything inherently wrong with such discussions, it makes me wonder why, especially recently, some of us have suffered from the accusation that we did not have the right to talk to McDonald, Hynek, Hall, the Lorenzens, etc., in the same way. He is not simply discussing with these persons what the project is doing and "methodology," but asking *how* "we" should best write the conclusions he has come to. I am impressed by the fact that it seems as if he is trying very hard to say as little as possible in the final report, but to say it in the most negative way possible. I do not think it is an unfair conclusion on our part to say that Bob is misrepresenting us, and that we have very definite grounds for feeling that our work, as represented by him, might not have much impact or importance. (I quote Dave Saunders when I say that Bob's suggestion that we could use footnotes for any minority opinions evoked Dave's response, "What do we do? Footnote the title?")

In the same sense that Bob has sought support from "outsiders" on what he is going to write in the final report, why is it unreasonable for us, feeling that what we said made very little dent on Bob's prejudged opinions, also to seek support from "outsiders"? Actually, the allegation of "prejudging" isn't the most important issue here. Even if he had not prejudged the problem, which I feel he did, his methods of arriving at his conclusions would still deserve a good deal of criticism.

I admit to a great deal of involvement with persons outside the project. I don't feel that talking to any of the people mentioned earlier (McDonald, Hynek, etc.) was wrong, except in the sense that sometimes it was easy to let frustrations show and possibly, in terms of the exact letter of office ethics, I was not always as tactful as I could have been. I was at the meeting in Denver in early December in which Saunders, Levine, McDonald, and Hynek got together to discuss the possibilities of action that might help to keep the study of UFOs going. All that was discussed there was totally independent of the C.U. project and would not have been a threat to the project in any way. In addition, I know that at that meeting McDonald received a copy of Bob's memorandum written to Deans Manning and Archer, although he knew the contents of it long before then. The substance of the memorandum, no matter the circumstances under which it was written or the fact that it was an internal piece of information written before the project started, serves mainly to substantiate to me the allegation that Bob has not done an honest job of representing himself in the UFO study.

In regard to McDonald's letter to Bob, in which he alludes several times to information that the "project members" have given him, I was present at a conversation in Tucson in March of 1967 where Bob, in the presence of both Jim Wadsworth and me, literally gave McDonald most of the information he could have asked for if he wanted to be antagonistic to the project. At that time Bob said: Condon does not have to look at cases, that is what we (including himself) are doing. In response to McDonald's question about the number of scientists we had on the project (both from the point of view of specialties and man-hours), Bob replied that we had as many as we needed and McDonald didn't need to tell us how to run the project. In addition, Bob said that you were not spending much time on the project, but that you shouldn't have to. (I believe that he thought he could do the job. However, I think the whole staff would agree that we did need you.) Therefore, I find it hard to feel now that, if McDonald is right in his accusations that our project has not been run well or even scientifically, we are much more guilty than Bob in transmitting that information to him. Dave and Norm were told that what they did was inexcusable, that they should not have communicated written information to someone outside the project. For this they were fired. I'm saying here that if giving McDonald the memorandum was a breach of office ethics, that Bob and the rest of us have breached that ethic, too. Bob asked me recently to see that some of the reports of cases that the C.U. project has investigated be sent to Dr. Menzel. These cases certainly contain confidential information and it is hard for me to draw the line between sending case information and sending internal memoranda—at least in principle. In any event, because of this, the project is now left with only two or three senior staff.

You have said that what Dave and Norm have done to the University in terms of ramifications that would make the University look bad is despicable. I think what they did in that sense is directly comparable to publishing our final report as a commercial book that would bring profit to the University. I can't imagine that the University would appear in a very good light if it looked as if we wanted to make money on this project. Yet this is what

Bob has been doing the past week—contacting publishers to see who will publish our report.

I think it is understandable that Dave and Norm felt an allegiance to something more than the UFO project as it existed. Up to their dismissal I felt it, too. And so did most of the others. After the last couple of days. I agree that I, and some others, have made a very tragic mistake in not coming to you long before this. But that is in retrospect and, at the time, I personally did not feel that you would have been as sympathetic to our feelings as you have been. Mistakenly or not, we felt that Bob did represent you, that he did talk to you often, and that therefore you were well-informed on what he was doing and what our position was. At the meeting we had in September following the statements you made (albeit misquoted) in *The Rocky Mountain News* we felt that we had "said our piece," and that our dissension was fairly open. I think we expected that after you would spend more time trying to correct what was possibly an incorrect impression of you on our part. Moreover, earlier that day when we were discussing the problems your statements might cause your staff, Bob excused himself from the discussion on the grounds that if he took part in our conversation concerning displeasure over what you said, he would not be able "to go back to" the administration. I do not know what his staying at the meeting and returning to his job in Regent Hall had to do with each other, but it certainly was not a very tactful way to handle the situation and did not leave us with a very good interpretation of his position.

I think I've rambled long enough, Dr. Condon, and therefore I shall end by saying that I am resigning my position as administrative assistant to the UFO project. I greatly appreciate your listening to me Thursday as sympathetically as you did. It seems that all there is left to say is that what I have written in this letter is one of the hardest things I've ever had to do, and that if it weren't for the fact that I believe what I have said very strongly, I would never have said it.

Sincerely,

Mary Louise Armstrong

APPENDIX 4

Excerpt of a Letter from J. Allen Hynek to Colonel Raymond S. Sleeper

7 October 1968

SECTION A

Blue Book has been charged with two missions by AFR 80-17, both ostensibly of the same weight, since the regulations do not specify otherwise. They are: (1) to determine if the UFO is a possible threat to the United States, and (2) to use the scientific or technical data gained from study of UFO reports. Neither of these two missions is being adequately executed.

First, the only logical basis on which it can be stated that UFOs do not constitute a possible threat to the United States is that so far nothing has happened to the United States from that source. First, many reports are not investigated until weeks or even months after they are made; clearly, if hostility were ever intended, it would occur long before the report was investigated. (That is akin to having the Pearl Harbor radar warnings [which went unheeded] investigated three weeks after Pearl Harbor.) Nothing did occur, so it can be gathered that UFOs, whatever they may be, have not so far had hostile intent.

Second, many reports of potentially high intelligence value go unheeded by Blue Book. Examples: (a) [Extract from a classified document of reported sighting of 5 May, 1965, contents unclassified, classification refers to name, and location and mission of vessel.] "... leading signalman reported what he believed to be an aircraft. . . . When viewed through binoculars, three objects were sighted in close proximity to each other; one object was first magnitude, the other two were second magnitude. Objects were traveling at extremely high speeds, moving toward ship at undetermined altitude. At . . . four moving targets were detected on the . . . air search radar at ranges up to 22 miles and held up to six minutes. When over the ship the objects spread to circular formation directly overhead and remained there for approximately three minutes. This maneuver was observed both visually and by radar. The bright object which hovered off the starboard quarter made the larger presentation on the radar scope. The objects made several course changes during the sighting, confirmed visually and by radar, and were tracked at speeds in excess of 3000 (three thousand) knots. Challenges were made by IFF but not answered. After the three minute hovering maneuver, the objects moved in a southeasterly direction at an extremely high rate of speed. Above evolutions observed by CO, all bridge personnel and numerous hands topside."

This report was summarily evaluated by Blue Book as "Aircraft," and to the best of my knowledge was never further investigated. By what stretch of the imagination can we say that the sighting did not represent a "possible threat" to the United States? Only because nothing happened. Do we ascribe such incompetence to the officers of the ship, and to the CO, to have such a report submitted unless all witnesses were truly puzzled? Is it conceivable that these officers could not have recognized an aircraft had it had the trajectory, the apparent speed, and the maneuvers ascribable to aircraft? No mention is made in the report of even the possibility that ordinary aircraft were being observed. The very fact that IFF challenges went unanswered should have been a spur to further investigation. This implies enemy craft. But the report does not even suggest the possibility that these were ordinary enemy aircraft. The classified document in Blue Book files does not contain further technical data concerning the sighting itself. Should not the director of Blue Book have exhibited at least *some* curiosity about this sighting? Yet when I brought it up on more than one occasion, it was dismissed with boredom. It is cases like these (but not this one, for it was never made public), apart from the question of possible threat, that add fuel to the Air Force "cover-up" charges that have been made from time to time by the public. It is hard for the public to understand how a country whose military posture is so security geared could dismiss a case like this out-of-hand unless the military knew more than they were telling.

(b) Extract from unclassified report received at Headquarters USAF, from the US Air Force District Office in Saigon and transmitted to Blue Book on 26 May, 1967. The date of the sighting was 17 April, 1967, or *more than a month before the report was received at FTD.* If there was a possible threat, Blue Book surely would not have known it! Why did transmission to FTD take so long? But to the report itself:

Statement of a member of the 524th Military Intelligence Detachment, Saigon Field Office, 205/8 Vo Tanh, Saigon, Vietnam: "At approximately 0220 hours, 17 April 1967, I observed five (5) large, illuminated, oval-shaped objects, traveling in close formation and at a very high rate of speed across the sky. At that time I was on the roof of the Saigon Field Office of the 524th MI Detachment. . . . I first saw these objects near the horizon to my left and watched them cover the entire field of my vision in what I believe to be less than five (5) seconds. During that period of time, the objects traveled from where I first saw them, near the horizon to my left, passed almost directly over me at what seemed to be a very great height, and then moved out of sight behind a cloud formation at the horizon to my right. The sky was partly cloudy but at the time of the sighting, the area of the sky over which they traveled was very clear with the exception of a few small patches of scattered clouds, which they seemed to be above. As the objects passed over these clouds, they were obscured from my vision until they emerged on the other side. I also observed that, as they passed between my line of sight and a star, they covered the star and blocked out its light until they had passed. This indicated to me that the objects were not transparent. It was apparent that they were not any form of conventional aircraft due to their size, shape, rate of speed and the fact that they made no noise audible to me. Prior to the sighting of these objects, I had been observing various conventional aircraft, both propeller and jet powered, and there is no question in my mind that they were a great deal larger than any craft I have ever seen in the sky. They were also traveling at a rate of speed which I would estimate to be at least five times greater than any jet powered aircraft I have ever seen. They were too distant and traveling too fast for a detailed description to be possible. I was only able to see that they were definitely oval in shape and glowed a steady white. They seemed to be in a vertical attitude, rather than horizontal, in relation to the earth, and their formation slowly fluctuated as they passed. Approximately five (5) minutes after they passed out of sight, several jet powered aircraft, which seemed to be at high altitude and traveling very fast, came from my far right and to my back as I faced the same direction as when I had seen the ovals. They proceeded to the area where I had lost sight of the objects, and upon reaching that point, they turned to their right and pursued the same course as the objects I had previously sighted. These aircraft were not in a formed pattern, but were scattered. I have never held any opinion concerning unidentified flying objects. Neither have I ever seen any previously. However, I believe that these objects were space craft of some kind. I am convinced that they were not reflections, conventional aircraft, meteorites or planets."

Now the above was an official report to Blue Book from USAF Headquarters, yet the case is carried in Blue Book as "Information Only." No follow-up was made, and *no evaluation was attempted*, on the grounds, I believe, that it happened outside the continental limits of the United States. The fact that it happened in a very sensitive area seemed to be of no concern to the director of Blue Book! Yet Blue Book states that UFOs represent no threat to the security of the United States. On what grounds? Only that so far nothing has happened.

Is it conceivable that no one in the military structure of the United States paid any attention to this sighting or correlated it with other reported sightings like it? Is there no curiosity as to pattern, no scientific curiosity in Blue Book? Apparently not.

It must be pointed out that neither of these cases were shown to me by Blue Book personnel. I happened upon them by accident during one of my visits as I scanned through material lying on a desk, and not in the files; I am not permitted to peruse the files themselves. I have access to the files only when I request a specific case. But how can I request a specific case, to examine its possible scientific merits, if I don't know of its existence? I am certain, from past attitudes of Blue Book, that I would never have been shown these cases; fortunately I came upon them (and many others) only by accident. And, I might say at this point, that when I do request a case, and wish to have a copy of portions of an unclassified case, I am not permitted to make a copy on the xerox machine just a few steps away—even when I offered to furnish my own xerox material! I must request same through "Reproduction" and thus endure a wait of possibly several weeks before I get a few sheets which I could have had in a few minutes. My usefulness as a consultant is thus grossly impaired.

(c) For the last example please see Section G which deals with the unscientific and unbusinesslike attitude within Blue Book. The two cases already stated likewise apply equally well also under Section H, since clearly no basic attitude of scientific curiosity was exhibited by Blue Book personnel in these two cases, and their scientific consultant was not even apprised of the existence of the reports.

SECTION B

The staff of Blue Book, both in numbers and in scientific training, is grossly inadequate to perform the tasks assigned under APR 80-17, even were they of a mind to do so.

This conclusion will be amply supported by what follows in the remaining sections, but it is clear that in dealing with a phenomenon which has puzzled a great many people, a problem that truly demands an interdisciplinary approach, two officers who hold only bachelor's degrees in physics from lesser institutions of higher learning, do not constitute an adequate task force for this problem. Even, however, were the officers Nobel prize winners, they could not do justice to the many reports that come into the Blue Book office. One baffling case could keep a staff of investigators busy for days or even weeks; trying to do justice to two or three cases a day over and above the peripheral duties attached to the office (see Section E) is clearly impossible.

SECTION C

Blue Book suffers intermurally in that *a* talks to *b*, *b* talks to *c* and *c* talks to *a*. More recently, it has been just a matter of *a* and *b*, and often it appears only *b*;

i.e., only one person is concerned in the evaluation of a report, with no cross-check. Blue Book is a closed system. It has, so to speak, fallen victim to the closed loop type of operation, to its own propaganda. There has been little dialogue between Blue Book and the outside scientific world or between Blue Book and the various scientific facilities within the Air Force itself. There has been little cross-fertilization of ideas and little or no contact with other groups, particularly civilian engineering groups, that have expressed an interest in the problem. As consultant, I have probably received more correspondence from other scientists and engineers about UFOs than Blue Book has since the closed type of operation of Blue Book is well known to such people and it is known that only sterotyped PR type of answers will be given by Blue Book to others. I know of very little scientific correspondence in the Blue Book files; this is probably because scientists wish to correspond with persons of like training. It would be pointless, for instance, to query Blue Book on the scientific reasons for evaluating a given case, say, as caused by a temperature inversion: Blue Book has never availed itself of the meteorological know-how within the Air Force itself to determine just how much of an inversion is necessary to produce the effects reported by the witness, if at all. The approach has been qualitative rather than quantitative; a two degree inversion is accorded as much weight as a 10 degree inversion, and not once have I seen geometrical optics applied to ray tracing in a given case evaluated as having been caused by an inversion. The staff is not adequate for this type of work. I have recently asked the Chief Scientist (see Appendix A) to initiate a request of AFCRL to compute and furnish tables to Blue Book which would give the optical effects to be expected from temperature inversions of varying degrees of intensity.

Similarly, many astronomical evaluations have been made by Blue Book without consulting their scientific consultant (who is, after all, an astronomer) which have brought ridicule in the press. The midwest flap of reports of July 31–August 1, 1965 can be cited as an example.

SECTION D

The statistical methods employed by Blue Book are a travesty on the branch of mathematics known as Statistics. A chapter in a doctoral dissertation in

Northwestern University, soon to be published, deals specifically with this aspect, and I will later quote from it (Herbert Strentz, "A Study of Some Air Force Statistical Procedures in Recording and Reporting Data on UFO Investigations," included in "A SURVEY OF PRESS COVERAGE OF UFOs, 1947–1967, a doctoral thesis at the Medill School of Journalism, Northwestern University") and preface it with my own observations which, incidentally, I have repeatedly brought to the attention of the Blue Book staff but to no avail. I finally felt it pointless to continue to try to educate the staff on these matters.

In the evaluation of cases it has been the custom to employ the terms "possible" or "probable" as modifiers to a given evaluation; thus, "possible aircraft" or "probable meteor" are often used. However, in the year-end compilation of cases these modifiers are quietly and conveniently dropped. Thus "possible aircraft" becomes simply "aircraft" (the Redlands case, [see Sec. 1] will appear in the final tabulation for 1968 as "aircraft") and the public will be led to believe that there was no possible question involved but that some poor citizen or citizens had had "one too many," or simply had been overexcited or suggestible.

Now a statistician will tell us that the words "possible" and "probable" should carry some idea of percentage probability. How probable? 50% probable? Only 100% probable is certainty. I think we might find general agreement among statisticians that it would be fair to assign 50% probability to the case "probable aircraft" and perhaps 20% probability to the term "possible aircraft." Thus if at year's-end 200 cases have been classed as "aircraft" in the final tally, but 100 of these were "probable" aircraft and 100 were "possible aircraft" then the probability is that of the 200 cases only 50 + 20 = 70 were actually aircraft and that thus 130 may not have been aircraft at all! For what else does "possible" or "probable" mean other than one is not *sure* they were aircraft. But so ingrained is the hypothesis of the "deluded observer" in Blue Book thinking that any other possibility is *not examined for*. This is hardly the scientific method.

Another illogical and unscientific method of Blue Book is the following: from the year 1947 through 1966, Blue Book has placed 1,822 cases out of a total of 10,316 in the "insufficient information" category. I might point out that the decision to make that classification is entirely subjective, except that sometimes a rule of convenience is used. Thus I have found the following notation on a recent case: *"In accordance with present policy the sighting is being carried as Insufficient Data since it was not reported to the Air Force within thirty days"*! By what possible legerdemain or reasoning can a sighting reported forty days after occurrence and containing ample information possibly be classed as insufficient? Hardly science. I would flunk one of my students who perpetrated such a travesty on the scientific method.

"In accordance with present policy. . . ." Whose policy? Certainly the scientific consultant was never consulted about this, or as a matter of fact, on very little other policy.

To return to the general cases bearing the mark, "Insufficient data": it is most interesting to note that such cases are carried in the statistics as having been solved, as though giving a case the Insufficient Data label constituted solving it! Here again the public is misled. Over the twenty years, my personal statistics show that out of 10,137 cases, 557 are listed as Unidentified and 1822 as Insufficient Data. The Blue Book handout reports that only 5.4% of the cases remain Unidentified, conveniently forgetting that 1822 additional cases, or 17.6%, remain unexplained. The correct figure of unidentified should therefore be 23%! When in past years I remonstrated with Blue Book officers, I have been left with the feeling that, "This is the Air Force. We have all the answers, and who are you to suggest a change in our established ways?" In the face of such attitudes, I was as Czechoslovakia was to Russia; resistance would only have led to bloodshed, and I felt it beneath my dignity to argue such points with the insufficiently trained personnel traditionally assigned to Blue Book. During one long period, a sergeant with no scientific background other than in psychology was doing nearly all of the case evaluations (Sgt. Moody). I continued as consultant in the face of all this largely out of a desire to have access to data which someday I might be able to use in a more productive manner, partly out of a desire to monitor the UFO phenomenon, and partly out of a sense of responsibility to the continuity I had maintained with the project over the years.

I quote now directly from the doctoral dissertation mentioned earlier: "The problem was underscored in an October 6, 1958, Department of

Defense press release on Blue Book activity from July 1, 1957, through July 31, 1958. The release said, 'More than 84% of the reported UFO sightings were *definitely established* (emphasis added) as natural phenomena . . . or man-made objects.'" Not only had the *probably* and *possibly* labels been deleted from the statistics, but sightings previously considered only *possibly* explained were now "definitely established"—not because of further investigation, but because of bookkeeping procedures.

"Lt. Col. Hector Quintanilla . . . acknowledged that the 'definitely established' phrase was 'misleading.' Defending the general procedure, however, he asked, 'Where else would you put it (the probably-possibly explanation)? Too many categories would make the report too cumbersome.' He added that continuing the *probably-possibly* categories year after year would only result in more work for Blue Book and lead to more questions."

Now, I ask you, Commander, is that Science? Did Madame Curie worry because her work was "too cumbersome"? Or that a scientific procedure "would only result in more work"? I could rest my case right there about the non-scientific approach by the staff of Blue Book.

I continue to quote from the forthcoming doctoral dissertation. "The monthly stratified sample (Mr. Strentz is now speaking of how he did his statistics) was drawn from every other year, beginning with 1948, the first full year of the Air Force UFO inquiry. Three months were selected from each even-numbered year, 1948 through 1966—one month from January, February, or March, one month from May, June or July, and one month from October, November or December. This provided a cross-section of UFO and Blue Book staff activity. . . 1,034 cards (Project 10073 Record Cards, Form 329) were examined. The number of sightings recorded by Blue Book for the same months was 1,117. So, cards were available for more than 90% of the reports recorded during the sample months (no mention is made why this wasn't 100%). The 1,034 cards also represented 9% of the number of UFO reports recorded by the Air Force from 1948 through 1966—11,038."

Later in his work Mr. Strentz states: "As a matter of routine, Project Blue Book considered *Insufficient Data* and *probably* or *possibly* cases as 'solved' in that there was no further investigation and the reports were categorized."

(No scientist would consider an "Insufficient Evidence" case as "solved." These simply should not be included in the data.) That is one way of bringing up the high Blue Book score—false, unscientific, but a lovely number to parade for the PR boys.

The Strentz report continues: "As shown in Table 1 (not reproduced here) analysis of the summaries found that 270, or 24%, of the 1,117 UFO reports were 'unsolved' or 'doubtful.' The 270 were those reports classified as Insufficient Data or Unknown, the great majority being the former. From the individual cards, analysis showed that 538, or 51%, of the 1,034 cases were 'unsolved.' The 528 were cases classified *probably, possibly, Insufficient Data*, and *Unknown*."

Thus, by simply advancing the *probably, possibly* cases to "established" status, the bookkeeping improved the Blue Book investigatory capacity by reducing the number of "unsolved" cases from 51% to 24%. Further, by emphasizing only the Unknown cases, Department of Defense press releases dealt with "unsolved" cases not of 51% or 24% but of "less than 2 percent," "less than 1 percent," and "2.09 percent." Thus has Blue Book passed itself off as being scientific and made a nice showing before the PR boys. When this doctoral dissertation is published by the Medill School of Journalism of Northwestern University, the high scientific capacity and prowess of Blue Book in explaining all but two or so percent of cases will be disclosed—less than one-half of the cases submitted to Blue Book have been solved!

The section of Mr. Strentz's doctoral dissertation which deals with Blue Book statistics closes with the following words: "Most of the UFO reports appear in fact to be 'unsolved.' So why not recognize that they are, that it is often impossible to determine what it was that an individual said he saw in the sky? The statistical methodology employed by Blue Book appears to have resulted from (1) Air Force efforts to explain every single report of a UFO because (2) the Air Force has been saddled with the unwelcome burden of proving that UFOs do not exist. Perhaps the Air Force mission might be redefined to deal only with sightings which promise some scientific paydirt and not with every report of a moving or hovering light in the sky. Then, the Air Force and the press might have something to work with, other than misleading statistics."

To all of which I can add a hearty Amen. This leads logically to the next point, E.

SECTION E

There has been lack of attention to significant UFO cases, as judged by the scientific consultant and others, and too much time on routine cases which contain few information bits; too much time and effort are demanded of the Blue Book staff for peripheral tasks (public relations, answering letters about evaluation of old cases and answering requests for information from various and sundry sources). The Blue Book staff, unless greatly expanded, and if it is to execute a scientific mission, should concentrate on two or three significant cases per month (such cases to be decided upon by consultation with a scientific panel) with the end result a scientific report in detail on each case, published as a scientific report and available to the public. Cases chosen should *not* be those in which only one witness is concerned (except in very unusual circumstances) or cases in which lights are seen in the distance at night, or cases in which the witnesses are judged of low reliability and are unable to make articulate responses to questions. As scientific consultant to Blue Book I have long advanced a method of judging those cases worthy of attention: a two dimensional classification whereby a case is judged by its *Strangeness* and by the composite *Credibility* of the witnesses of the sighting. By *Strangeness* is meant a measure of the difficulty of honestly explaining the sighting by well known physical phenomena and principles; the composite final *Credibility* of the witnesses can, of course, only be determined by their past records, medical and social, and by whatever psychological tests it may be feasible to apply. An estimated credibility can be quickly judged, however, by simply noting the number of witnesses and the responsibility each carries in his daily life. Clearly, only cases rated high on the *Strangeness* and estimated *Credibility* scales need be considered. All told, Blue Book has wasted far too much time on cases of little significance, and in other areas, on time-wasting peripheral tasks.

Further, once a case has been classed as *Unidentified* or *Unknown* that is the end as far as Blue Book is concerned. In Science, the unknown, the unexplained, is the *start*, and not the end of inquiry. A scientist who finds something in his laboratory that he can't explain is no scientist if he labels it "unknown" and files it away and spends the rest of his time in routine matters. It is precisely the Unknowns that Blue Book should be concerned with, not making impressive (?) counts of how many people cannot properly identify a satellite or a meteor. That might be of some interest to a sociologist, but hardly to a physical scientist.

SECTION F

The information input to Blue Book is grossly inadequate and certainly the cause of much of the inefficiency within the Blue Book office itself. An impossible load is placed on Blue Book by the almost consistent failure of UFO officers at the local Air Bases to transmit adequate information to Blue Book, and, I might say, it was considerably worse in the long period before there were UFO officers so designated.

Many "information bits" of possible crucial value in the evaluation of a case are missing in the original report. I have seen so many that it is virtually nauseating. At best, the original UFO report as it comes in to Dayton is an intelligence-type report, and hardly a scientific report, but its content and value could be very greatly improved if the UFO officers at local Air Bases really took their jobs seriously. Many information bits which could have been obtained by conscientious interrogation by the UFO officer are omitted, thus throwing the burden upon Blue Book's already very small staff to reopen the interrogation to obtain the necessary information—sometimes of a most elementary and obvious sort, e.g., wind direction, angular sizes and speeds, details of trajectory, contrast of object with sky, availability of other witnesses, etc. A prime example of this is the Redlands, California, case, quoted below, see G, in which the blame must be placed almost entirely on the local officer, who sent so little information through to Blue Book that the latter failed to recognize its significance.

It would appear that Blue Book has never been given enough authority to "throw a case back into the teeth" of the local interrogator and to demand immediate further information. If the military has anything, it has, because of its command structure, the means whereby such information can be demanded, and not merely politely asked for and the request allowed to be disregarded. The upgrading of original data is one of the most pressing needs within Blue Book. "We are smelting a very low grade ore."

SECTION G

The basic attitude within Blue Book is unscientific in that a working hypothesis has been adopted which colors and determines the approach to the problem. We state a theorem:

> For any given reported UFO case, if taken by itself and without respect and regard to similarities to other UFO cases in this and other countries, it is always possible to adduce a possible natural explanation if one operates solely on the hypothesis that all UFO reports, a priori, because of the nature of the world as we presently understand it, must result from well known, accepted causes.

Corollary:

> It is impossible for Blue Book to evaluate a UFO report as anything other than a misidentification of a natural object or phenomenon, a hoax, or a hallucination.

(The classification "unidentified" does not constitute an evaluation.)

The essence of the scientific method is that the investigator must not adopt a preconceived idea or conclusion, he must not select those bits of data which favor his hypothesis and overlook those that go against it. The salient scientific error perpetrated by Blue Book is portrayed in the above theorem. So certain is Blue Book of its working hypothesis that it reminds one of the doctor who was so certain that all abdominal swellings were the result of tumors that he failed to recognize that his patient was pregnant.

Let me choose just one example from a great many possible to illustrate the above charge, but one that illustrates well the gross lack of rigor in the scientific methodology of Blue Book.

I choose the incident at Redlands, California, of February 4, 1968, a recent case which was investigated by no one at Blue Book, superficially by a member of Norton AFB, and for a total of three months by Dr. Philip

Seff, professor of geology, Dr. Reinhold Krantz, professor of chemistry, Dr. Judson Sanderson, professor of mathematics, and artist John Brownfield, professor of art (who drew an artist's conception from the descriptions given independently by the witnesses and whose composite painting was verified by the witnesses), all of the University of Redlands. It is of interest to note that no one at Blue Book has seen fit to contact these investigators and discuss their investigation at least over the phone.

The case itself concerns the reported sighting by some twenty observers of an object with seven lights on the bottom, which appeared as jets, and a row of eight to ten lights on top which were alternating in color. The object was reported to have proceeded at a low altitude (estimated about 300 feet) in a northeasterly direction for about a mile, to have come to a stop and to have hovered briefly, jerked forward, hovered again, then wavered to the northwest, gained altitude, and then to have shot off to the northwest with a strong burst of speed. It was under observation for about five minutes. The object was estimated to have been at least 50 feet in diameter. The estimates of 300 feet altitude and 50 feet in diameter must be considered jointly; only the apparent diameter can be judged, of course, but on the assumption of a given distance the estimate of 50 feet was arrived at. Clearly, if the object had been several miles away, the unchanged apparent diameter would lead to an unbelievably large object. For these reasons these estimates cannot be summarily dismissed.

You will undoubtedly be interested to know that Blue Book classified this object as "probably aircraft." How this was arrived at with no investigation is, of course, a striking example of methodology of Blue Book. Norton AFB reported that March AFB radar painted no unusual targets (ignoring completely the fact that an object at 300 feet altitude would have been missed by this radar) and that a light plane had landed at Tri-City airport at 19:15 PST, whereas a check of the police blotter and of all witnesses agreed that the sighting could *not* have occurred earlier than 19:20. Further, a check made by the university professors, (but apparently not even thought of by Blue Book) with the authorities at the airfield showed that the plane was coming in from Los Angeles and *never approached closer than six miles to the city of Redlands* and therefore never passed over the city of

Redlands, whereas all witnesses agree that it was actually close over the city. The plane which landed (which Blue Book did not think to inquire about) was a Bonanza single engine propeller aircraft which the professors took the trouble to examine while in its hangar at the airfield. [The Redlands case is the sole subject of a book now in production by David Branch and Robert Klin, entitled *Inquiry at Redlands.*]

The discrepancy between what was reported and the Blue Book evaluation is so great as to be laughable. The law, further, states that planes cannot fly lower than 1000 feet over Redlands. It appears inconceivable that twenty or so witnesses would misidentify a light, single engine plane, several miles away, as a brilliantly lighted, unconventional aircraft at 300 feet that jerked, hovered, and sped away, and went straight up into the overcast.

But no weight at all was given by Blue Book in this case, as in a great many other cases to which I can attest, to the *possibility* that something strange might have been going on. In a most unscientific manner, every item was slanted and biased in favour of the Blue Book working hypothesis. It was assumed, against good evidence, (1) that the time of observation was in error (2) that an unusual, low flying object would have been picked up by radar (despite the fact that low flying planes in test exercises have succeeded in getting through our defense radar cover) (3) that all witnesses could not distinguish between six miles and 300 feet (4) that all witnesses could not distinguish between a light, single engine plane, which could hardly carry a battery of extremely bright lights above and below, and (5) that witnesses could not distinguish between the smooth maneuvers of a plane in a landing pattern miles away and hovering, jerky, and fast motions of the object reportedly viewed. Finally, (6) it was assumed that the professors involved had not the intelligence to recognize for themselves (having been over the ground and having "reenacted the crime" so to speak), the possibility of the witnesses having misinterpreted a plane in a landing pattern, and have been individually wrong on the time, the place, the motion, the brightness, and the number of lights. And, over and above this is another tacit assumption, however politely hidden, that not only the witnesses but the professors were demented or incompetent, for only under such an assumption could one seriously advance the evaluation of "probable aircraft."

It should be remembered that Blue Book made no on-the-spot or telephonic investigation at all, Norton AFB spent less than two man-days on the investigation, such as it was, for when all but one of the witnesses was asked whether they had been interviewed by an Air Force representative, the answer was negative.

Now, if it should turn out that all witnesses and private investigators were incompetent, deluded, and psychotic, and that it was indeed a plane that caused the sighting, that conclusion can be reached only by sheer intuition, and not by the "scientific" investigation conducted by Blue Book. In any court of law it would be unthinkable to allow a prosecuting attorney to distort, deny, and disregard the testimony of several witnesses to a crime in order to prove the guilt of the defendant. And in science we like to think that we employ far more rigorous, objective, unbiased methods than are employed in a courtroom where emotional bias can and does creep in.

SECTION H

Inadequate use has been made of the Blue Book scientific consultant and the scientific liaison he represents. He has only limited access to files in that he must first know of a case before he can ask for the relevant files. Often he has been unaware that a certain case existed until he either accidentally stumbled upon it or it was brought to his attention by outside agencies.

In all of his twenty years association with Blue Book, only now has he been asked to evaluate its methodology. He has now been asked to recommend means for "product improvement." The product at present has little public value, the product image is poor, the product does not inspire public confidence, and the method of processing the raw material, packaging the product, and distributing it violates many principles of good business. Incidentally, the product is not selling either.

In view of the limited staff of Blue Book, limited in numbers and in scientific training, it may seem hopeless to accomplish anything worthwhile, and I am tempted to recommend that Blue Book be abolished as essentially worthless and the problem turned over to competent scientific personnel. For the UFO problem very probably will vanish, in this or other countries, with or without Blue Book. The AFR 80-17 clearly states that the objectives of Blue Book are twofold: "to determine if the UFO is a possible threat to the United States and to use the scientific or technical data gained from a study of UFO reports." The key phrase here is "from study of UFO reports." I must ask, "what study?" Should you say that that is my business, I must reply that I am but one person, whose time is already committed nearly fully to academic matters. As a *consultant* I can do my best to guide and advise, but except in special circumstances, that is all I can do. However, I have strongly advised in the past on how the study of UFO reports might proceed so as to obtain whatever they might contain of scientific value. And that method is *not* the method that was employed in the Redlands case and many others. As I have often said to students, "If you think you know the answer in advance, it isn't research." To study UFO reports means to consider them as research data and to handle them as a mature scientist would handle data he obtains by observation in nature or in the laboratory. Granted that UFO reports are fragmentary and often subjective; so are the reports received by intelligence teams, sociologists and poll takers. Yet they manage to do something with them. But when Blue Book received the previously quoted UFO report (see Section A) from a member of the 524th Military Intelligence Detachment operating in Saigon, a trained observer, of completely unconventional objects which covered horizon to horizon in five seconds although flying higher than the clouds, and blocking out stars as they flew past, Blue Book refused my strong request that this be investigated on the grounds (1) that it was outside the US and hence no concern of theirs, and (2) there was probably nothing to the report in the first place!

Blue Book also refused to act upon my request that a report made by Dr. Roger Woodbury, Associate Director of MIT's Instrumentation Laboratory (sighting of January 14, 1966) be fully investigated by local intelligence officers, who could certainly have established whether any special scientific exercise was being carried on at that time from any of the local air bases. The scientific apathy shown by the officers of Blue Book in this and many other cases has ceased to amaze me. When one has a report from a highly placed scientist in one of the nation's greatest scientific laboratories, one should pay attention. The scientists who have produced Polaris should be reckoned with when in all seriousness they report an unusual happening.

Summing up: the methodology of Blue Book is unscientific in that no scientist would test only for a preconceived hypothesis and rule out summarily even the possibility of another hypothesis; they would manifest scientific curiosity about the matters in hand; they would attempt to find patterns in data rather than handling each datum as though it existed in a vacuum. In case after case Blue Book, for instance, has dismissed a case because the local air base reported that no aircraft were in the area. In that event, argues Blue Book, the observer obviously must have been deluded. The proper scientific approach would, of course, be to seek a solution that is consistent with the basic data of the report and not with the working hypothesis.

SECTION I

I must point out that I have made recommendations in the past for "product improvement," but which went unheeded. I refer to the paper (AFCIN-4E2x) entitled, "ATIC UFO Investigation Capability" and signed by Col. Evans. This grew out of hearings held in Washington July 13–15, 1960.

Present on July 15 were: Mr. Robert Smart, Armed Services Committee, Mr. Spencer Beresford, Mr. Richard Hines and Mr. Frank Hammil, House Science and Astronautics Committee, Mr. John Warner, CIA (Assistant for Legislative Liaison to Mr. Allen Dulles), Mr. Richard Payne, CIA (Tech. Advisor), Mr. John McLaughlin, Adm. Assistant to Secretary Air Force, M/Gen. A. H. Leuhman and B/Gen. E. B. LeBailly, SAFOI, B/Gen. Kingsley and Col. James McKee, SAFLL, L/Col. Sullivan, AFCIN-Pla, L/Col. Tacker, SAFOI-3d, Maj. J. Boland, SAFLL, Maj. Robert Friend and myself.

Had the recommendations arrived at in those meetings (recommendations which I strongly supported) been implemented, Project Blue Book would today probably have a decent scientific record rather than being the letter-writing, filing, and monitoring agency that it in fact is.

These recommendations were:

1. Blue Book should have the capability to investigate those cases which give an indication of having high intelligence or scientific potential, and also those which generate an unusual amount of public interest. In making this recommendation, Mr. Smart stated that the investigative capability of Air Force Bases is limited to routine cases and that the Air Force should have both the numbers and capability to conduct the UFO operation. This was taken to mean that Blue Book should investigate outstanding cases and an indication of the high priority assigned to this was that for an interim period the financing of this program was authorized to be made directly from the Office of the Secretary of the Air Force. (This was never done—in a letter dated September 21, 1960, Blue Book was notified that no additional funds or personnel could be authorized and that investigations must be made with currently available personnel and resources.) This, coupled with the fact that purely routine (largely PR) duties occupy a large part of the present small and totally inadequate staff, and that the present staff has not been chosen for its scientific background (as judged by scientific training, scientific publication record, or any of the standard methods current among universities in the selection of their science faculties) but rather, it seems, by expediency of whatever officer happened to be available or who gave assurance that "the boat would not be rocked" by too much insistence that a proper scientific job be done (I refer particularly to previous officers in charge of Blue Book, most of whom seemed to be sitting around waiting for their retirement, and one notable one who spent a great deal of his time planning his brokerage office after retirement) or by one who would be intelligent enough to properly use the excellent scientific facilities of the Air Force, such as those of the Air Force Cambridge Laboratories, where radar and meteorological experts are available. For instance, in the long history of Blue Book, never once was Cambridge asked to calculate whether the inversions to which a sighting was attributed were actually sufficient,

quantitatively, to account for the UFO sighting. (The word "inversion" has indeed become a catch-all around Blue Book—given that a 3° inversion existed at 6,000 feet, this has been used to explain a sighting made by an aircraft at 15,000 feet!)

2. Mr. Smart requested that summaries of all significant cases be forwarded to his office. (To the best of my knowledge this has never been done.)
3. Project Blue Book office must have immediate mobility and capability to investigate cases of importance. (This recommendation was based largely on my insistence to the Committee that Blue Book invariably got "scooped" by civilian organizations in the investigation of cases. Time and again NICAP or APRO had interviewed the witness before the local Air Force men had, and I gathered from many witnesses that the civilian interrogations often were more thorough than those of the Air Force. I also pointed out the need for upgrading the data. Time and again the reports from local Air Bases constituted a waste of teletype time, as illustrated by one famous instance when the TWX carried two pages of addresses and the message, "Just another UFO." That was the content of the message!)

To function properly, Blue Book should have a sufficiently high-ranking officer in charge who could command that Air Force regulations be carried out at the local air base level not only to the letter of the Regulation but to their spirit also. I have personally been told repeatedly of the ridicule meted out at the local level, and of the superficial and often cavalier methods of investigations. I knew for a fact that obviously relevant information bits have been omitted in cases where the interrogating officer just apparently did not care enough to ask relevant questions, which would have served to establish some quantitative idea as to the angular speed, apparent brightnesses, the kinematics of the reported object, and where certainly no attempt was made to seek out other witnesses. (To this point Colonel Quintanilla has informed me that "Blue Book is not an investigative agency"—how in the name of common sense can a scientific job be done without investigation! Investigation is the very lifeblood of science.)

Thus we see that long before we can speak of "product improvement" we must seek means of improving the raw material from which our product eventually stems.

I also pointed out to the Smart Committee that when certain original data are lacking, as they very often were, immediate telephonic contact with key witnesses was essential—not two or three months later, but within hours of the receipt of the TWX. The first order of business at Blue Book should be the immediate scrutiny of an incoming report, to decide whether it is "significant" in the terms already specified and, if so, to decide at once what additional information is needed and to proceed to get it immediately—calling in the Scientific Consultant then, and not weeks later, to ask his help in gathering information. Despite the fact that my time is limited, I do have an excellent scientific staff that could be employed from time to time to obtain such information. I refer particularly to Mr. William Powers, systems engineer, who has on many occasions proved his ability to interrogate witnesses in a meaningful and confidence-inspiring manner. Mr. Fred Beckman, of the University of Chicago, has also been of great help to me, and entirely on a voluntary basis.

Unfortunately, the recommendations made and applauded in Washington were never implemented. The result was that with the limited staff, their many duties, an ingrained feeling that the whole subject was worthless, and the statement by the Project Director that "we are not an investigative agency," Blue Book is a routine, dull, uninspired operation—so much so that it would be psychologically impossible for me to be associated with it physically on a daily basis. (Indeed, Lt. Marano has complained to me that his strong desire to be transferred is due to the fact that he hasn't been able to use what scientific training he has in his job.)

The Air Force should finally recognize the UFO phenomenon as a global, scientific problem of possible great potential and attempt to fulfill the second part of its two-fold mission by asking that Blue Book be aided by a scientific panel drawn from the various already existing scientific missions within the Air Force as well as outside scientific groups; and that this panel start from where the Condon Committee will have left off.

It might be best, however, in the long run, to ask that the Blue Book second mission be transferred out of the Air Force entirely and given to a civilian group comprised of capable scientists from various disciplines, since the problem approach is undoubtedly interdisciplinary.

Which of the two available paths is followed must be decided largely by yourself and the exigencies of the situation. I remain ready to assist you in both reaching and implementing your decision.

J. ALLEN HYNEK, DIRECTOR
LINDHEIMER ASTRONOMICAL RESEARCH CENTER
NORTHWESTERN UNIVERSITY
EVANSTON, ILL.

Chapter Notes

CHAPTER ONE

1. Schrodinger, Erwin: *Nature and the Greeks.* p. 55.
2. A personal communication from Thomas Goudge to the author.
3. A personal communication from Dr. Fred Whipple to the author.
4 Barlay, Stephen: *The Search for Air Safety.* Wm. Morrow & Co., Inc.: New York, 1970. p. 145.
5. *Journal of the Medical Society of New Jersey.* Vol. 66. August, 1969, pp. 460-64.

CHAPTER TWO

1. See Appendix 1, NL-13.
2. See Appendix 1, CEI-2.
3. From an interview with a woman from Kenora, Canada, about her sighting of May 30, 1969. This case is not listed in Appendix 1 because it had only one witness.
4. Sighting of May 24, 1965. Report is not included in Appendix 1 because author had no personal contact with the reporter or the investigator.
5. The sighting took place on June 8, 1966, in Kansas, Ohio.

CHAPTER THREE

1. From a letter to the commanding general, Wright-Patterson Air Force Base.
2. From a personal letter to the author reporting a UFO sighting.

3. See Appendix 1, CEI-3.
4. Taken from a letter to the author reporting a UFO sighting.
5. See Appendix 1, DD-13.

CHAPTER FIVE

1. Observers of Selected Nocturnal Lights Events

OCCUPATION	NUMBER
Air Control operations	8
Teenagers	4
Children	4
Housewives	3
Police officers	2
Antiques dealers	2
Air force crew	2
Service station attendants	2
Butcher	1
Laborer	1
MIT graduate student	1
Royal Canadian Air Force telecommunications operator	1
Associate Director MIT Physical Lab	1
Air force major	1
Air force 1st lieutenant	1
Medical doctor	1
US Naval security member	1
Civilian pilot	1
Shop man	1
Unknown	3
Total	41

2. This statement was made by an Air Force intelligence officer who investigated the case.
3. These instructions are fully covered in Joint Army-Navy-Air Force Proceedings (JANAP-146E).

4. The following is the full text of the letter I wrote to *Physics Today:*

> More than a year has passed since the Air Force formally closed its Project Blue Book, which acted as a national center for the receipt of reports of certain types of strange phenomena more commonly known as UFOs.
>
> As consultant to that project for many years I am aware that neither the closing of Blue Book nor the Condon report has laid the UFO problem to rest, and a number of my scientific colleagues and I have become concerned lest data of potential scientific value be lost for want of a reporting center. As evidence that the subject is still very much alive under the covers, I can cite not only my own personal mail, which continues to contain UFO reports from reputable persons, but also news-clipping services. The latter show an almost complete absence of UFO reports from urban dailies but a continued spate of UFO reports from small-town newspapers, where the editor is either less sophisticated or less prone to be influenced by officialdom, or where he may have knowledge of the source of the UFO reports.
>
> It has been my estimate over the past twenty years that for every UFO report made there were at least ten that went unreported. Evidence for this comes from the Gallup Poll, the many UFO reports I subsequently learned of that were not reported to the Air Force, and from my own queries. There has always been a great reluctance to report in the face of almost certain ridicule. It would seem that the more trained and sophisticated the observers, the less prone they are to report unless they could be assured of anonymity as well as respect for their report.
>
> Accordingly, in order that material of potential scientific value not be lost, and in order that persons, particularly those with scientific training and experience, can submit a UFO report without fear of ridicule and publicity, my colleagues and I, all associated with universities, hereby offer to act as a receipt center for UFO reports that otherwise would almost certainly be lost to science. I will be personally responsible that the data so submitted will be treated seriously and that no embarrassment to the sender will result. Names, for instance, will be immediately disassociated from the report and not used without specific written permission of the originator.

It may be of interest to note, in passing, that over the years I have been the recipient of UFO reports from many highly trained technical people and scientists. It is a gross but popular misconception that UFO reports spring from "ding-alings." A study of the record shows that such persons are almost entirely absent. The address to which UFO reports may be sent is: J. Allen Hynek, Chairman, Department of Astronomy, Northwestern University, Evanston, Illinois 60201."

CHAPTER SIX

1. Observers of Selected Daylight Sighting Events

OCCUPATION	NUMBER
Army artillery trainees	12
Teenagers	6
Civilian pilots	4
Farmers	4
Children	5
Technicians	3
Research engineers	2
Prospectors	2
Scientific balloon observers	2
Housewives	4
Air force base personnel	2
B-17 pilot	1
Astronomer	1
Meteorological engineer	1
Commercial pilot	1
Physiotherapist and former USAF pilot	1
Army veteran, now university student	1
Secretary	1
Owner of baseball team	1

Security policeman	1
Unknown	3
Total	58

2. Baker. "Observational Evidence of Anomalistic Phenomena." Journal of the Astronomical Sciences, 15, 31 (1968).
3. *The original negatives were returned to the owner, who then submitted them to the Condon committee. In the committee report it was stated that these photographs "have no probative value."*
4. *Mr. Fred Beckman, a colleague who has frequently assisted me in UFO photographic matters, made the tests on the negatives in question.*
5. *I mention this circumstance only to show that this report would never have been made except for this condition and would have remained, I suspect, in the large reservoir of latent reports. The original observers had no intention of reporting the incident officially. In my many years as a UFO investigator I have repeatedly encountered an overwhelming reticence to report officially, especially to the police or to the Air Force. Many letters sent to me carry the specific injunction not to transmit the information contained to the Air Force. In this particular instance, however, the report was transmitted directly to Dayton, where I came upon it in the course of my routine monitoring of reports.*

CHAPTER SEVEN

1. "Unidentified Flying Objects," Hearing by Committee on Armed Services of the House of Representatives, 89th Congress, 2nd Session (April 5, 1966), the Honorable L. Mendel Rivers (chairman of the committee) presiding, p. 6073.

   MR. SCHWEIKER: . . . have any of the unexplained objects been sighted on radar. I thought you said no to that just a couple of minutes ago.

   MAJOR QUINTANILLA: That is correct. We have no radar cases which are unexplained.

2. Observers of Selected Radar-Visual UFO Reports

OCCUPATION	NUMBER
Radar operators	15
Airport control operators	7
Ship's crew members	6
Military pilots	6
Commercial pilots	5
Military airmen	3
Ship's bridge personnel	3
Private pilots	3
Private plane passengers	2
Airmen (Second Class)	2
Airmen (Third Class)	2
Airman (First Class)	1
Ship's master	1
Able seaman	1
Ordinary seaman	1
Third mate	1
Commanding officer (ship)	1
Director of operations–bomber wing	1
Total	61

3. *Flying Saucer Review.* Vol. 16, No. 2. March/April, 1970, pp. 9-17.
4. *Astronautics and Aeronautics.* July and September, 1971.

CHAPTER EIGHT

1. Observers of Close Encounters of the First Kind

OCCUPATION	NUMBER
Housewives	8
Teenagers	8
Patrolmen and police officers	7
Adult males, occupation unknown	2

Cabinet maker	1
College student	1
Waitress	1
Ex-nurse	1
Naval trainee	1
Elementary school teacher	1
Chemistry teacher	1
School principal	1
Former naval officer (now real estate broker)	1
Graduate student in anthropology	1
President, small airline	1
Businessman	1
Night watchman	1
Instrument maker	1
Farmhand	1
Clerk	1
Total	41

2. UFO Report No. 66-26 A/B, NICAP, Massachusetts Investigating Committee.
3. There is good evidence that tracking stations, both visual and radar, and amateur observing groups, such as Moonwatch stations, have observed UFOs but have not reported them because it was not politically expedient to do so.
4. Records of the Smithsonian Astrophysical Observatory show that neither of the bright satellites, Echo I and Echo II, was in the sky over Portage County at the time. Nor were any of the three Pegasus satellites visible at that time. Even if these had been, however, their brightness was five to ten times less than that of the Echo satellites, and their orbital inclination was so low that they would have been seen only to the south.
5. The taped interviews, with more than ten persons directly involved with the episode, represent some sixty hours of taping; this stands as an exemplary case of UFO investigation and should be made public.

CHAPTER NINE

1. Observers of Close Encounters of the Second Kind

OCCUPATION	NUMBER
Housewives	18
Teenaged girls	17
Teenaged boys	10
Adult males (occupation unknown)	8
Employees and family members of Canadian fishing resort	6
Businessmen	5
Engineers	4
Pilots	3
Farmers	4
Police officers	2
Boys (ages 6-10)	3
Truck drivers	2
Senior highway designer	1
Roofer	1
Schoolteacher and former Air Force flight attendant	1
Supervisor in mail order house	1
Collection manager, finance company	1
Chief of technical services, Air France	1
Beekeeper	1
Professional artist	1
Painter	1
Hairdresser	1
Total	92

2. I would hope that readers, pilots and others, who might heretofore have been reluctant to admit their experiences will be encouraged to submit an account of the experience to the author with the same understanding.

3. I am indebted for this report to Raymond Fowler, whose meticulous and detailed investigations of many New England cases far exceed in completeness the investigations of Blue Book. He has regularly sent me copies of his reports and has given me permission to quote from them.
4. Correspondence with officials in the Levelland area has shown that such radio contact did not exist at that time.
5. Levelland *Sun-News*. November 6 and 7, 1957.
6. I am indebted to the National Investigations Committee for Aerial Phenomena (NICAP) for material in addition to that in the Blue Book files, for whom James Lee, of Abilene, Texas, carried out a personal investigation of the Levelland occurrences. He encountered a report that two grain combines, each with two engines, that had been operating in Petit, Texas (about fifteen miles northwest of Levelland), were silenced by the passing of a glowing UFO.
7. I am indebted to Ted Phillips, Jr., an assiduous independent investigator with whom I have worked closely. He has specialized in recording, cataloging, and investigating this one relatively narrow but highly important aspect of the subject for many of the cases used in this chapter.
8. Bernier, publisher of *UFO-Info*. Seattle, Washington. February 12, 1966.

CHAPTER TEN

1. Bowen, Charles, ed. *Humanoids*. Henry Regnery, Chicago; Vallee, Jacques. *Passport to Magonia*, Henry Regnery, Chicago; Bowen, Charles. *Flying Saucer Review*. London, now in its eighteenth year of publication; *Phenomenes Spatiaux*, Paris; and *Lumieres dans la Nuit*, Paris.
2. UFO Landings with and without Occupants (omitting years 1952, 1967, 1968, 1969, for which only partial Blue Book records are not available)

	VALLEE CATALOG	REPORTED TO BLUE BOOK
Landings in the United States	190	48
Landings in the US with occupants	65	12
All landings (worldwide)	546	–
All landings with occupants (worldwide)	223	–

3. Blue Book Evaluations of Landing Cases

	ALL LANDINGS	LANDINGS WITH OCCUPANTS
Insufficient Data	8	1
Hoax	4	2
Psychological	4	2
Unreliable reports	1	1
Balloons	1	0
Fire	1	0
Aircraft	2	0
Ground light	2	2
Hallucination	1	1
Moon and Venus	1	0
Meteor	1	0
Birds	1	0
Satellite	1	0
Inconsistent data	1	
Radar inversion	0	1
Unidentified	7	2
Total	36	12

4. Crotwell, Norman E. *A Report on Papuan Unidentified Flying Objects.* Anglican Mission, Papua, New Guinea.
5. *Ibid.*
6. Ledwith's account is reproduced with his permission.
7. This refers to a hypothesis proposed by D. I. Warren, *Science,* November 6, 1970, pp. 599–603, titled "Status Inconsistency Theory and Flying Saucer Sightings," in which he maintains that UFO reports are more apt to come from people whose economic status is not consistent with their intellectual capacity and training, for example, a poorly trained person occupying a relatively high economic and social status, or vice versa.

CHAPTER ELEVEN

1. I was an associate member of that panel but was not invited to participate in all the sessions. In one session I attended, the famous Tremontian,

Utah, and the Great Falls, Montana, movies (well known to all who have followed the UFO saga) were shown and dismissed as seagulls and aircraft, respectively. The panel, of course, did not have the benefit of the detailed analysis of the Great Falls case ("Observational Evidence of Anomalistic Phenomena," *Journal of Astronautical Sciences,* Vol. XV, No. 1, 1968, pp. 31–6) made by Dr. M. L. Baker, carried out under the auspices of the Douglas Aircraft Company, by whom Dr. Baker was then employed. In his paper Dr. Baker concludes, ". . .the images cannot be explained by any presently known natural phenomena."

I was dissatisfied even then with what seemed to me a most cursory examination of the data and by the set minds implied by the panel's lack of curiosity and desire to delve deeper into the subject. For by 1953 there already existed many hundreds of cases of high S-P (it was a far cry from the early Project Sign cases); the panel examined about a dozen. I was not asked to sign the report of the panel, nor would I have done so had I been asked.

2. Menkello, F. G. "Quantitative Aspects of Mirages." Report No. 6112. Menkello is a first lieutenant, USAF, Environmental Technical Applications Center. "It is easy to show that the 'air lenses' and 'strong inversions' postulated by Gordon and Menzel, among others, would need temperatures of several thousand degrees Kelvin in order to cause the mirages attributed to them."
3. Ruppelt, Edward. *Report on Unidentified Flying Objects,* p. 80.
4. *Ibid.*, p. 81.
5. *Ibid.*, p. 81.
6. *Ibid.*, p. 82.
7. *Ibid.*, p. 83.
8. *Ibid.*, p. 88.
9. Kepler, the German astronomer who, unable to garner data himself, used data obtained throughout the years by the Danish astronomer Tycho Brahe, who, in turn, had no idea what to do with his excellent data. Kepler and Brahe had many arguments, yet Kepler knew that he needed those data in order to construct his theory of planetary motion. So he bided his time.

10. *Application of Electronic Data Processing Techniques to Unusual Aerial Phenomena: Organization and Development of an Inquiry System.* Submitted by J. Allen Hynek. July, 1966.
11. To that end, a number of my scientific colleagues and I at Northwestern University have agreed to act as a receipt center for UFO reports, especially from persons with scientific and technical backgrounds. It is important that data of potential scientific value not be lost.

CHAPTER TWELVE

1. The membership of the committee and an illuminating history of its two-year existence can be found in *UFOs? Yes!* by David Saunders and Roger Harkins (Signet Book No. 3754). The constituency of the committee without an illuminating history can also be found in the Condon Report, "Scientific Study of Unidentified Flying Objects." Both books are "must" reading for serious readers of the actions of the Condon group.
2. To test the Condon recommendation to government funding agencies, I submitted two serious research proposals, one to the National Aeronautics and Space Administration and the other to the National Science Foundation. Both were summarily rejected not because of scientific unworthiness (or so the rejection letters stated) but because of lack of funds.
3. Hearing by Committee on Armed Services of the House of Representatives. 89th Congress, April 5, 1966, No. 55.
4. Saunders, *op. cit.*, p. 141
5. Saunders, *op. cit.*, Chapter 15, "Condon's Favorite Cases."
6. "Unusual Aerial Phenomena." *Journal of the Optical Society of America*, April, 1953.
7. Powers, W. T. "A Critique of the Condon Report." Refused publication in *Science* in 1969.
8. Saunders, *op. cit.*, Chapters 19 and 20.
9. Fuller, John G. "Flying Saucer Fiasco." *Look*, May 14, 1968.
10. Mrs. Armstrong has graciously given me permission to quote her letter in the interests of the historical record.

CHAPTER THIRTEEN

1. *Flying Saucer Review.* Special Issue No. 4, August, 1971, pp. 57-64.
2. *Journal of Astronautics and Aeronautics.* Vol. 9, No. 7, July, 1971, p. 66.
3. November, 1970.
4. Private communication from Julian Hennessey, from a personal letter to Sir John Langford-Holt, M.P. : "In the normal course of events UFO records would remain closed to public scrutiny until they became available under the usual rules at the end of thirty years. However, if a major scientific organization of high standing had strong reasons for obtaining access to our records, then its application would be considered on its merits."

EPILOGUE

1. Hoyle, Fred. *Of Men and Galaxies.* Seattle: University of Washington Press, 1964, p. 47.

About the Author

Josef Allen Hynek (1910–1986) was an American astronomer, professor, and ufologist. He is best remembered for his UFO research. Hynek acted as scientific advisor to UFO studies undertaken by the US Air Force under three consecutive projects: Project Sign (1947–1949), Project Grudge (1949–1952), and Project Blue Book (1952–1969). In later years, he conducted his own independent UFO research, developing the Close Encounter classification system. He is widely considered the father of the concept of scientific analysis of both reports and especially trace evidence purportedly left by UFOs.

To Our Readers

MUFON BOOKS

The mission of MUFON BOOKS, an imprint of Red Wheel Weiser, is to publish reasoned and credible thought by recognized authorities; authorities who specialize in exploring the outer limits of the universe and the possibilities of life beyond our planet.

ABOUT MUFON

The Mutual UFO Network (*www.MUFON.com*) was formed by concerned scientists and academic researchers in 1969 for the specific purpose of applying scientific methods to the serious study of UFO sightings and reported human/alien interactions. MUFON's mission is "The Scientific Study of UFOs for the Benefit of Humanity" with the intent to unveil and disclose credible information free of distortion, censorship, and lies, and prepare the public for possible implications.

ABOUT RED WHEEL/WEISER

Red Wheel/Weiser (*www.redwheelweiser.com*) specializes in "Books to Live By" for seekers, believers, and practitioners. We publish in the areas of lifestyle, body/mind/spirit, and alternative thought across our imprints, including Conari Press, Weiser Books, and Career Press.